NETWORK CODING

Edited by **Mohammad A. Matin**

Network Coding

http://dx.doi.org/10.5772/intechopen.72225

Edited by Mohammad A. Matin

Contributors

Josu Bilbao, Goiuri Peralta, Raul Gomez, Pedro Crespo, Syed Ali Hassan, Rafay Iqbal Ansari, Muhammad Arslan Aslam, Chrysostomos Chrysostomou, Charoula Mitsolidou, Christos Vagionas, Dimitris Tsiokos, Nikos Pleros, Amalia Miliou, Abel Ajibesin, Mohammad Abdul Matin

Notice

Statements and opinions expressed in the chapters are these of the individual contributors and not necessarily those of the editors or publisher. No responsibility is accepted for the accuracy of information contained in the published chapters. The publisher assumes no responsibility for any damage or injury to persons or property arising out of the use of any materials, instructions, methods or ideas contained in the book.

First published in London, United Kingdom, 2018 by IntechOpen

IntechOpen is the global imprint of INTECHOPEN LIMITED, registered in England and Wales, registration number: 11086078, The Shard, 25th floor, 32 London Bridge Street

London, SE19SG – United Kingdom

Printed in Croatia

British Library Cataloguing-in-Publication Data

A catalogue record for this book is available from the British Library

Additional hard copies can be obtained from orders@intechopen.com

Network Coding, Edited by Mohammad A. Matin

p. cm.

Print ISBN 978-1-78923-614-9

Online ISBN 978-1-78923-615-6

We are IntechOpen,
the world's leading publisher of
Open Access books
Built by scientists, for scientists

3,650+
Open access books available

114,000+
International authors and editors

118M+
Downloads

151
Countries delivered to

Our authors are among the

Top 1%
most cited scientists

12.2%
Contributors from top 500 universities

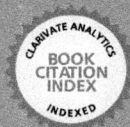

CLARIVATE ANALYTICS
BOOK
CITATION
INDEX
INDEXED

WEB OF SCIENCE™

Selection of our books indexed in the Book Citation Index
in Web of Science™ Core Collection (BKCI)

Interested in publishing with us?
Contact book.department@intechopen.com

Numbers displayed above are based on latest data collected.
For more information visit www.intechopen.com

Meet the editor

Dr. Mohammad A. Matin currently works as an associate professor at the ECE Department, North South University, Bangladesh. He obtained his BSc degree in Electrical and Electronic Engineering from BUET (Bangladesh), his MSc degree in Digital Communication from Loughborough University, UK, and his PhD degree in Wireless Communication from Newcastle University, UK. He has published over 90 refereed journals and conference papers and is the author/editor of 12 academic books, such as *Communication Systems for Electrical Engineers* (Springer, 2018), *Spectrum Access and Management for Cognitive Radio Networks* (Springer, 2016), *Coding for MIMO-OFDM in Future Wireless Systems* (Springer, 2015), and *Advances in Sensor Networks Research (Nova Publishers, USA, 2014)*. He is currently on the editorial board of several international journals, such as *IEEE Communications Magazine, IET Wireless Sensor Systems*, and so on.

Contents

Preface

Network coding (NC) is a novel approach to optimize the flow of data by performing coding operations across a network to increase the capacity of the network and improve its throughput and robustness. In recent years, a significant amount of research has been performed to explore the impact of NC in different scenarios and enhance network performance.

This book attempts to expose the most recent important observation on network coding research. The essential issues of NC as well as some important future research directions are illustrated in this book. Therefore, it is hoped that it will serve as a comprehensive reference for graduate students who wish to enhance their knowledge of NC for networked systems.

Mohammad A. Matin
North South University
Bangladesh

Chapter 1

Introductory Chapter: Network Coding

Mohammad Abdul Matin

Additional information is available at the end of the chapter

http://dx.doi.org/10.5772/intechopen.79423

1. Introduction

Network coding is a novel approach that allows nodes in the network to perform coding operation at the packet level. In particular, network coding represents a powerful approach to protect data from losses due to link disconnections and can also allow exploiting the combination of multiple links to deliver data to users with the possibility of recoding at intermediate nodes. This phenomenon will reduce the information congestion at some nodes or links which will improve the network information flow such as to increase network throughput and robustness.

2. Outline of research contributions

This book attempts to present cutting-edge research in the field of network coding.

In *"Digital All-Optical Physical-layer Network Coding,"* the authors present the concept of digital All-Optical Physical-layer Network Coding (AOPNC) for mm-wave fiber-wireless signals modulated with up to 2.5 Gb/s OOK data, focusing on digital encoding schemes that are based on optical XOR logical gates. The encoding operation is performed on-the-fly at the Central Office (CO), and the resulting packet is broadcasted at the end users, where the electrical decoding takes place. The AOPNC scheme in principle can be applied also in RoF networks employing other phase modulation formats, such as DPSK-SCM and dual polarization (DP)–DQPSK-SCM modulation techniques.

In *"Network Coding for Distributed Antenna Systems,"* the authors explore virtual MIMO-assisted distributed antenna system (DAS) and network coding (NC) to improve the performance of networks. An analysis is presented to provide design insights that could help in identifying the network parameters to achieve the desired QoS. The results highlight the advantages of employing NC in VMIMO-assisted DAS.

In *"Bringing the cloud to the fog for Industry 4.0,"* the authors focus on the benefits and open challenges of the Industrial IoT (IIoT) architecture together with the implementation of network coding techniques. The IIoT architectures require low-latency communications as well as guaranteed reliability to allow the performance of on-premise advanced cloud analytics for time-critical IIoT applications, i.e., bringing the cloud to the fog. This chapter also describes the communication process across the different levels of the architecture based on network coding.

In *"Efficient Frontier and Benchmarking for Energy Multicast in Wireless Network Coding,"* a network coding algorithm is studied, and its performance is investigated for the data envelopment analysis (DEA). The DEA methodology is necessary because coded packet is not fully efficient technique for energy efficiency. The DEA framework allows network administrators to evaluate the technical efficiency rather than averages and standard deviation and determine how the inefficient wireless networks will attain a targeted efficiency frontier. The presented system model is based on frontier analysis that is consisting of several models including envelopment and benchmarking. These models are considered for evaluating the technical efficiency of multicast energy and performing the benchmark in wireless networks nodes without sacrificing the overall network performance. The author's aim is to achieve economic efficiency by ensuring that wireless networks are multicast at the targeted energy rather than average energy.

3. Conclusion

Recently, the field of network coding (NC) has attracted intense research focus for its potential in providing enhanced network throughput and reduced network congestions. However, it is challenging to incorporate network coding, into the existing network architecture. This book provides few current research efforts which are supplemented with extensive references to enable researchers for further investigation of network coding applied to communications in wireless and wired networks.

Author details

Mohammad Abdul Matin

Address all correspondence to: matin.mnt@gmail.com

Department of Electrical and Computer Engineering, North South University, Dhaka, Bangladesh

Chapter 2

Digital All-Optical Physical-Layer Network Coding

Charikleia Mitsolidou, Chris Vagionas,
Dimitris Tsiokos, Nikos Pleros and Amalia Miliou

Additional information is available at the end of the chapter

http://dx.doi.org/10.5772/intechopen.75908

Abstract

Network coding (NC) has recently attracted intense research focus for its potential to provide network throughput enhancements, security and reduced network congestions, improving in this way the overall network performance without requiring additional resources. In this chapter, the all-optical physical-layer network coding (AOPNC) technique is presented, focusing on digital encoding schemes that are based on optical XOR logical gates. It is also discussed how digital AOPNC can be implemented between subcarrier-modulated (SCM) optical signals in radio-over-fiber (RoF) networks, circumventing the enhanced complexity arising by the use of SCM signals and the asynchrony that might exist between the data arriving at the encoding unit. AOPNC demonstrations are described for simple on/off keyed (OOK)-SCM data signals, as well as for more sophisticated higher-order phase modulation formats aiming to further improve spectrum efficiency and transmission capacity.

Keywords: all-optical physical-layer network coding (AOPNC), radio over fiber (ROF), millimeter wave communication, optical logic gates, semiconductor optical amplifier-Mach Zehnder interferometer (SOA-MZI)

1. Introduction

The explosive data traffic growth in combination with the increasing use of smart mobile devices has created the need for high-throughput wireless access networks at Gb/s scale [1, 2]. In this context, radio-over-fiber (RoF) networks have recently stepped in as a promising solution to satisfy this demand by seamlessly converging the ubiquity and mobility of the "last-meter" wireless networks with the high capacity of backhaul optical networks [3, 4]. Up to now, there are several RoF signal generation and modulation schemes [5, 6] as well as advanced functionalities related to end-user mobility, hand-off schemes [7] and to network coding (NC) in RoF networks [8, 9]. Network coding (NC) has drawn significant scientific attention due to its capability to provide network throughput enhancements, security and low latency by improving the capacity resource utilization and enabling bidirectional data transport [10].

Until now, NC has been demonstrated separately for purely wireless networks [11] and passive optical networks (PONs) [12–14]. Regarding wireless networks, NC is performed at the relay by using conventional electronic processing of data, while most of the NC demonstrations in PONs rely on optical-electrical-optical (OEO) conversion before the encoding operation, resulting in further complexity at the central office (CO) and additional latencies to the overall communication [13, 15]. RoF technology was introduced as the technology that can merge wireless and optical functionalities via remote antenna units (RAUs) to deliver seamless communication between the wireless user and the CO. In order to comply with the requirements imposed by the trend of optical-wireless convergence, RoF technology should be able to implement unified NC between the optical network and the wireless user. Network coding concepts satisfying the earlier requirement are expected to enhance the overall network efficiency when implemented directly in the optical domain at the CO side.

However, till now, most of the optical PHY-layer NC (OPNC) demonstrations target the encoding of baseband optical signals transmitted in wired optical links. The cross-gain modulation (XGM) and cross-phase modulation (XPM) phenomena in semiconductor optical amplifiers (SOAs) and SOA-Mach Zehnder interferometers (SOA-MZIs) have been exploited to perform the XOR-based network coding between baseband on-off keyed (OOK) data signals [16, 17], while the four-wave mixing (FWM) phenomenon in SOA- [18] and highly-nonlinear fiber (HNLF)-based XOR gates [19] has been employed for the encoding of differential phase shift keyed (DPSK) signals. Although all the aforementioned demonstrations provide performance enhancements in wired optical links, they are not suitable for RoF networks due their incapability to deal with the sub-carrier modulation (SCM) formats of the RoF signals and the possible asynchrony between the data arriving at the NC-encoding unit.

Up to now, only a limited number of physical-layer OPNC demonstrations are compatible with RoF networks [8, 9, 20]. These schemes utilize orthogonal polarization multiplexing and optical power addition yet favoring analog physical-layer NC in order to cope with the increased coding complexity associated with SCM RoF data signals. Recently, a digital AOPNC scheme for RoF networks that can perform a bitwise XOR function between OOK-SCM data has been demonstrated [21]. Moreover, the current trend in RoF networks is moving toward the use of phase modulation formats and high-order modulation [22, 23], aiming to improve the spectrum efficiency and transmission capacity. Therefore, digital AOPNC schemes compatible with DPSK- and differential quadrature phase shift keyed (DQPSK) phase-formatted SCM signals have been recently demonstrated for future radio-over-fiber (RoF) networks [24, 25].

In this chapter, a digital AOPNC demonstration for up to 2.5 Gb/s OOK data modulated over a 10 GHz SC is presented. The proposed experimental setup uses a SOA-MZI XOR gate as the all-optical encoder between two SCM-OOK data streams, exploiting the low-pass filtering response of a SOA-MZI [26] in order to process the data envelope while discarding the

10 GHz SC. For validation purposes, the optical encoding operation is further evaluated by deploying a second MZI-XOR gate to decode the signal and retrieve the original information.

To extend the use of this concept in more sophisticated modulation formats and in mm-wave communications [6, 7, 27], a digital AOPNC scheme for 60 GHz SCM DQPSK signals was also investigated with physical-layer simulations for a bit rate equal to 4Gb/s. The all-optical encoder, residing in the central office (CO), consists of a delay interferometer (DI) stage for the DQPSK-to-OOK conversion, a stage of SOA-MZI-based OOK-XOR gates followed by the SOA-MZI-based phase regenerator [28, 29] that forms the phase-encoded signal. A remote 60 GHz oscillator wavelength feeder is also incorporated in the NC unit to generate an additional wavelength spaced by 60 GHz to the encoded signal wavelength for allowing up-conversion through the beating at the receiver site [30]. An electrical XOR decoder was used in the simulations, emulating the decoding operation that would normally be performed at the wireless users' site, assuming wireless transmission by the RAU to the wireless user.

The rest of the chapter is organized as follows: Section 2 presents the proposed AOPNC-RoF-based conceptual scheme; Section 3 includes the setup and the results for the coding between two OOK-SCM signals; Section 4 presents the AOPNC setup and results for the DQPSK-SCM signals. Finally, conclusions are addressed in Section 5.

2. Concept overview

Figure 1 illustrates the proposed AOPNC-based RoF network, comprising the NC unit at the CO and the two wireless end users communicated with the respective RAUs, which in turn are connected with the CO via a fiber. Users A and B transmit their packets to the RAUs, where they are converted to SCM optical streams through the modulation of the laser diodes (LDs), coupled together and forwarded to the CO through the optical media. The physical layer all-optical XOR-encoding operation is performed on the fly at the CO and the resultant signal is broadcasted back to both RAUs, de-multiplexed from the uplink traffic and converted from optical-to-electrical (o/e) by means of a photodiode (PD). The electrical encoded signal is transmitted from each RAU to the respective wireless user in

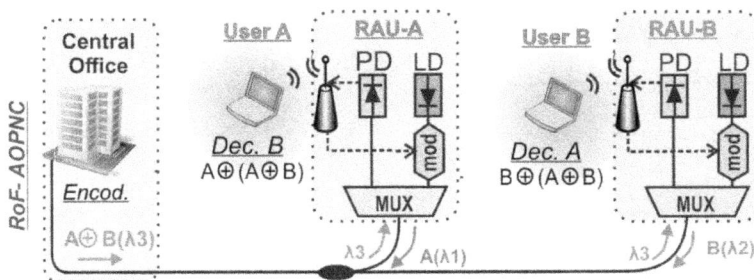

Figure 1. All-optical digital network coding (NC) scheme at the central office of RoF networks.

order to be decoded. Each user recovers the bit sequence originating from the other user by performing a second XOR function between its own locally stored data and the received NC-encoded data. In this way, the AOPNC technique implemented at the CO is capable of encoding information coming from users located in different cells, which is not feasible with current wireless NC schemes.

Figure 2(a) depicts the frame scheduling in an RoF network where no network coding is employed, as well as the frame scheduling in an RoF-NC-based communication scheme. In both cases, it is considered that for the uplink, RAU-A and B transmit data packets modulated on the wavelengths $\lambda 1$ and $\lambda 2$, respectively, while for the downlink traffic, the CO uses $\lambda 3$. When network coding is not employed, RAUs transmit Data A and B to the CO during the first timeslot, while the CO receives both packets and forwards packet from user A to B and packet from user B to A, using two successive timeslots. When network coding is applied in an RoF network, then the CO encodes and broadcasts the NC-encoded packet to both users in a single timeslot, occupied only two timeslots for the uplink and downlink traffic. This concept exploits the pre-amble and post-amble of the frames [31] for packet order resolving in case of non-synchronized packets. **Figure 2(b)** shows the encoding and decoding operation when packets from user A and B reach the encoding unit at the CO bit level synchronized or with a time delay equal to Δt (asynchronous operation). In the case of synchronous operation, data from user A and B are digitally encoded by means of a SOA-MZI XOR gate and broadcasted to both users. Each user recovers the packet of the other user by performing a second bitwise XOR operation between the received encoded signal and its own data. Considering that the end users are wireless clients that may reside at different distances from their respective RAUs resulting even to sub-bit time mismatch between the two packets reaching the CO, it is critical for the AOPNC scheme to operate also for asynchronous data.

As shown in **Figure 2(b),** during asynchronous operation, a sub-bit optical pulse with duration equal to Δt is generated in the encoding packet. However, the final decoded packets from the second XOR operation can still be correctly recovered by both end users through a second XOR operation, without any data loss.

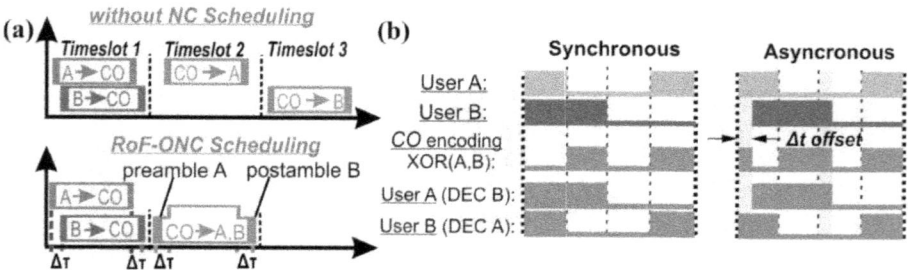

Figure 2. (a) Frame scheduling for RoF network: Without network coding and with the proposed AOPNC scheme and (b) conceptual encoding and decoding operation for synchronous and asynchronous data with a time offset of a sub-bit delay Δt.

3. Digital all-optical physical-layer network coding for OOK-SC signals

3.1. Experimental setup

Figure 3 shows the experimental configuration, exploiting two SOA-MZI gates, the first for the encoding process at the CO and the second for decoding XOR operation that in a realistic RoF is performed at the end user. The continuous wave (CW) signals at $\lambda1 = 1549.8$ nm, $\lambda2 = 1553.1$ nm and $\lambda3 = 1553.6$ nm were emitted by three tunable LDs (TLDs), multiplexed by an array waveguide grating (AWG) and modulated by an electro-absorption modulator (EAM) by a 10 GHz electrical clock signal for the SC generation. The output of the EAM was amplified by an erbium-doped fiber amplifier (EDFA) and de-multiplexed by means of an AWG. Signals $\lambda1$ and $\lambda2$ were further OOK modulated by two LiNbO$_3$ modulators, which were driven by a programmable pattern generator (PPG) loaded with 2.5 Gb/s NRZ 2^7-1 pseudo random bit sequence (PRBS), so as to form the SCM-OOK uplink data A and B signals. These uplink signals were transmitted through spools of single-mode fibers (SMF) with lengths of 3.9 km and 4 km, emulating the uplink connections between the RAUs and CO. An optical delay line (ODL) was employed at the branch carrying the data B stream, in order to enable bit-level synchronization during synchronous operation and insert time offsets at the asynchronous operation. Stream $\lambda3$ was not modulated with data in order to emulate the SC produced at the CO and used for the downlink traffic. Variable optical attenuators (VOAs) and polarization controllers (PCs) were also employed for power regulation and polarization adjustments, respectively.

The data streams ($\lambda1$, $\lambda2$) were injected into the control ports A and D, while $\lambda3$-SC was applied at the port C of the SOA-MZI XOR gate 1. The input power levels were measured, 800 μW(-1 dBm) for $\lambda1$, $\lambda2$ and 400 μW(-4 dBm) for $\lambda3$. The SOA-MZI was biased in such a way so that port G acts as the switching port. Hence, when only one of the two data is present, a π shift between the two SOA-MZI branch signal constituents is obtained by cross-phase modulation (XPM) and $\lambda3$ emerges with a logical "1" at the output port G. Otherwise, when both data signals are equally present, then $\lambda3$ bears a logical "0", confirming the implementation of an all-optical XOR gate by means of an SOA-MZI. The output port of the first XOR encoding gate was filtered and launched into port D of the second SOA-MZI XOR gate, which

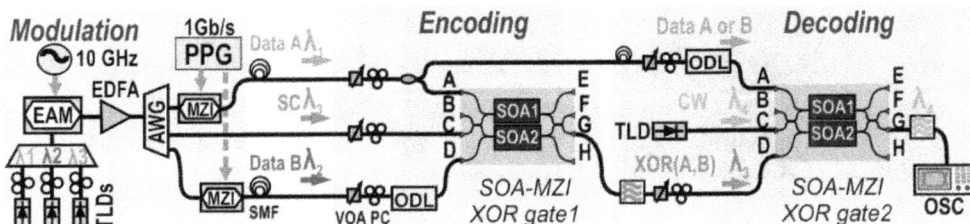

Figure 3. Experimental setup of the proposed AOPNC scheme comprising the SC-modulated data generation, the encoding and the decoding process.

acts as a decoder. A part of data A was connected with port A for decoding data B, while a CW signal launches port C ($\lambda 4$ = 1548.4 nm). Equivalently, when data A is decoded, data B is connected to port A of the second SOA-MZI. The input power levels were measured, −1 dBm for the control signals $\lambda 1$, $\lambda 3$ and − 4 dBm for $\lambda 4$. SMF and ODL were used for pattern and bit-level synchronization. The output of the XOR2 gate was filtered by a 0.65 nm filter and monitored by an optical sampling oscilloscope (OSC).

The SOA-MZIs featured two 1600 μm long hybrid-integrated SOAs, both operating at a moderate current value of 180 mA. The SOA gain recovery value at driving conditions was 180 ps for both SOA-MZI XOR gates, significantly longer than the 100 ps period of the SC signal. In this way, the SOA-MZI response of both XOR gates is turned into a low-pass filtering [26], neglecting the high-speed sub-carrier of the optical control data A and B signals but correctly processing their data envelopes.

3.2. Experimental results for synchronous operation

Figure 4 presents the experimental results obtained for the synchronous encoding and decoding operations between two 2.5 Gb/s data streams. **Figure 4(a)** and **(b)** shows the input time traces of data A and B, respectively, while the XOR stream at the output of the first SOA-MZI is shown in **Figure 4(c)**. The encoded stream features a logical "1" bit value when data A and B correspond to different bit values, while it is equal to "0" when the two data have the same logical bit value. **Figure 4(d)** and **(e)** illustrates the decoded data A and B at the output of the second SOA-MZI, after the XOR operation between the encoded signal and either the data B or data A, respectively. It should be noted that the decoded streams featured sub-bit dips

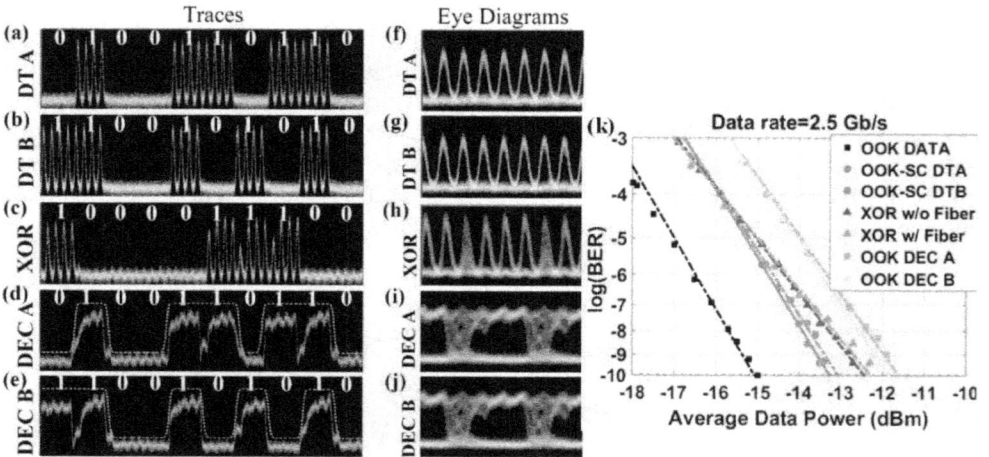

Figure 4. Experimental results for 2.5 Gb/s data. Traces (400 ps/div): (a) data a, b) data B, (c) encoded XOR, (d) decoded data a and (e) data B. Eye diagrams (80 ps/div): (f) data a, (g) data B, (h) encoded XOR, (i) decoded data a and (j) decoded data B. (k) BER curves for the OOK data without SC, the OOK-SC data, the encoded XOR with/without fiber after the users and the decoded output OOK data.

between successive "1"s, resulting from the XOR operation when logical "1"s are generated by the transition from differential phase $+\pi$ between the SOA-MZI branches to $-\pi$. However, the initial data patterns were retrieved successfully, confirming the successful decoded process at 2.5 Gb/s. **Figure 4(f)** and **(g)** depicts the eye diagrams of the input OOK-SCM data A and B, respectively. **Figure 4(h)** shows the eye diagram of the XOR-encoded signal at the output of the first SOA-MZI, exhibiting an extinction ratio (ER) of 9.6 dB and an amplitude modulation (AM) of 0.5 dB. **Figure 4(i)** and **(j)** illustrates the eye diagrams of the decoded data A and B, respectively, exiting the second SOA-MZI XOR gate. Both eye diagrams exhibit an ER equal to 8.2 dB and an AM equal to 1.2 dB.

The successful encoding and decoding operations were also verified with the aid of bit error rate (BER) measurements. **Figure 4(k)** shows the BER curves carried out at various stages of the system. The BER curves reveal error-free operations for both decoded data signals, having a power penalty equal to 3.2 dB when compared with the initial OOK data signals at BER = 10^{-9}. This power penalty is partially attributed to the SC modulation of data A and B, which introduces a power penalty of 2 dB. BER curves for the encoded XOR stream exiting the first SOA-MZI were also carried out for two cases: when the SCM-OOK DATA are directly inserted to the encoder without the use of extra fiber and when fiber spools of 3.9 km and 4 km are inserted between the OOK-SC data A and B and the first XOR gate in order to emulate the uplink connection. The power penalty between these two BER curves (with and without fiber) is negligible. By comparing the XOR curves with the OOK SCM data, the power penalty at 10^{-9} is approximately 0.7 dB.

3.3. Experimental results for asynchronous operation

Possible asynchrony between the two data streams reaching the encoding unit was also examined by introducing various sub-bit temporal delays at data B, as may potentially be introduced by mobile wireless users. **Figure 5(a)** and **(b)** illustrates the input traces of OOK-SC data A and the delayed-by-100-ps (0.25 τ_{bit}) data B, while **Figure 5(c)** depicts the encoded XOR trace. Although short "parasitic" pulses of 0.25 bit duration appear at the encoded XOR stream, by implementing the decoding XOR function between the encoded XOR and the delayed data B, the trace of the decoded data A can be successfully retrieved, as shown in **Figure 5(d)**. **Figure 5(e)** shows that data B was also correctly decoded by the XOR operation between the encoded stream and data A, after a time delay of 0.25 bit duration.

Similar results were obtained for the asynchronous encoding and decoding of NC operation where data B is delayed by a sub-bit time offset equal to 200ps (0.5 τ_{bit}), as shown in **Figure 5(f)–(j)**. **Figure 5(f)** and **(g)** illustrates the SCM-OOK data A and B, respectively, reaching the encoding XOR gate with a time offset of 200 ps. **Figure 5(h)** shows the encoded XOR trace exiting the first SOA-MZI, while **Figure 5(i)** and **(j)** shows the decoded data A and B after the second XOR operation between the encoded pattern and the initial data pattern of data B or A, respectively. In this operation, both data A and B were successfully retrieved, with decoded data B having a delay equal to the time offset. The highlighted insets magnify the traces in a specific part of the stream where the asynchrony can be observed and the XOR signal appears to have "parasitic" pulses equal to the respective time offsets.

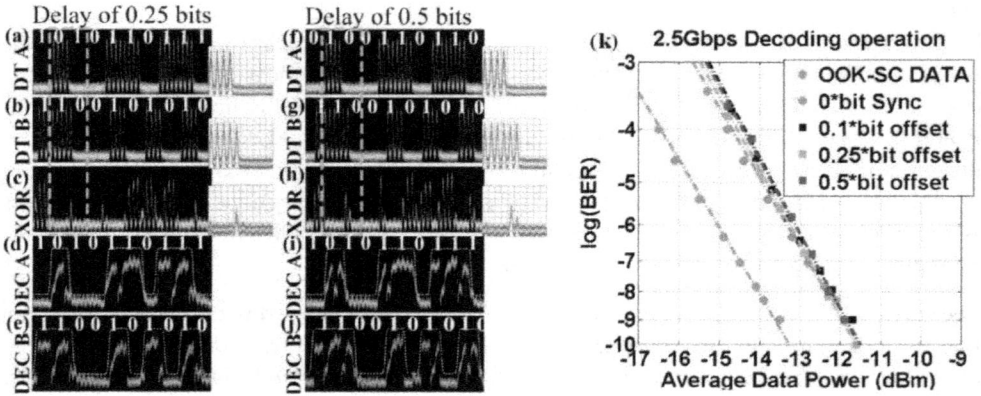

Figure 5. Experimental time traces of asynchronous 2.5 Gb/s operation for two different time offsets. Time traces (400 ps/div): For time offset equal to $0.25*\tau_{bit}$: (a) data a, (b) data B delayed by 0.25 bit, (c) encoded XOR signal, (d) decoded data a and (e) data B. For time offset equal to $0.5*\tau_{bit}$: (f) data a, (g) data B delayed by 0.5 bit, (h) encoded XOR signal, (i) decoded data a, (j) data B. Magnified insets highlight time offset. (k) BER curves of the decoded data signal for various time offsets between the input OOK-SC data.

The successful asynchronous operation was also evaluated with BER measurements, by inserting various relative delays between the two data and measuring the error rate for the down-converted decoded signal. **Figure 5(k)** shows the BER curves versus the average received power for time offsets equal to 0 τ_{bit}, 0.1 τ_{bit}, 0.25 τ_{bit} and 0.5 τ_{bit}, revealing error-free operation at 10^{-9} for all these cases. No additional power penalty when comparing the synchronous with the asynchronous operation was observed. This fact indicates that the performance of the proposed system remains functional even in the case of non-synchronized packets. The power penalty of the decoded stream with respect to the original down-converted OOK-SCM data is approximately 1.5 dB.

4. Digital all-optical physical-layer network coding for DQPSK-SC signals

4.1. Simulated setup

Figure 6 illustrates the setup employed in order to evaluate the proposed AOPNC scheme in an RoF network that uses DQPSK-SCM data signals. The data A and B streams are modulated employing the DQPSK format in the wireless users' transmitter (Tx) and transmitted to the RAUs for electrical-to-optical conversion. The CWs at $\lambda 1 = 1551$ nm (RAU A) and $\lambda 2 = 1553$ nm (RAU B), exiting the respective laser diodes (LDs), are modulated by the RF DQPSK signals via the MZMs. These signals are transmitted through a 4 km spool of SMFs, amplified by EDFAs and multiplexed by an AWG before entering the encoder in the CO premises. In the NC encoding unit, the data signals are inserted in two delay interferometers (DIs) in order to recover the OOK-u and -v complementary components. The differential optical phase between

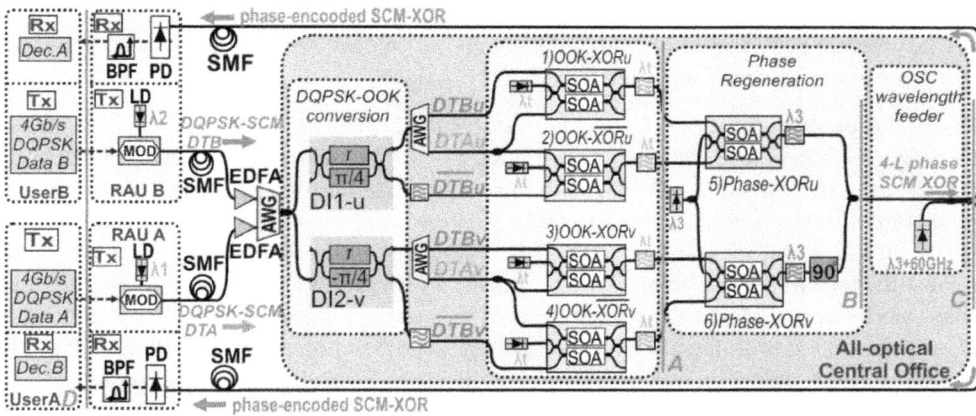

Figure 6. Simulated setup of the proposed AOPNC scheme comprising the two users, the respective remote antenna units (RAUs) and the all-optical encoding unit at the central office. Stages A–D are the stages where the OOK-XOR (XOR), the four-level phase-formatted XOR and the four-level phase-formatted optical and electrical SCM-XOR signals are generated, respectively.

DI1-u and DI2-v arms is set to 45 and −45° so as to recover the u- and v-constituents of data A and B, respectively. The u and v components correspond to the PRBS sequences before the differential encoding to I and Q bits as it is explained by Vorreau et al. [28].

The upper output port of each DI is connected with an AWG that de-multiplexes Data A and B signals, while the output of the DI's lower port is filtered by a band pass filter (BPF) with a center frequency equal to $\lambda 2$ that keeps the OOK-SCM $\overline{\text{DataB}}$ and ignores the OOK-SCM data A. The OOK-converted streams are inserted into the control ports of the OOK-XOR gates based on SOA-MZIs and a continues wave λt (temp)=1555nm was inserted in their probe port. The data A-u and B-u are injected into the first SOA-MZI so as the OOK-XORu signal to be generated at the output switching port. The data A and $\overline{\text{DataBu}}$-constituents are applied at the second SOA-MZI forming the $\text{OOK} - \overline{\text{XORu}}$ at the output of the switching port. Similarly, the OOK-XORv and $\text{OOK} - \overline{\text{XORv}}$ are obtained at the output ports of the third and fourth MZIs, respectively. All the OOK encoded signals are then filtered and driven to the phase regeneration stage where the amplitude-to-phase conversion is performed by means of two SOA-MZIs. Particularly, the OOK-XORu and $\text{OOK} - \overline{\text{XORu}}$ are inserted into the control port of the fifth SOA-MZI and a continues wave in $\lambda 3$=1557.36 nm was launched in the probe input port. The relative phase in the SOA arms was controlled by a high-level bit either in the OOK-XORu or the $\text{OOK} - \overline{\text{XORu}}$ arm, forming in this way the phase-encoded signal between data A-u and B-u components [28, 29]. Equivalently, the OOK-XORv and $\text{OOK} - \overline{\text{XORv}}$ streams are injected in the control arms of the lower SOA-MZI, forming the phase- and wavelength-($\lambda 3$)-converted XORv signal. The phase-formatted XORu and XORv output streams are filtered and recombined with a relative phase shift (PS) of 90° so as to form the four-level (4-L) phase-formatted XOR signal. Finally, the phase regenerator output at $\lambda 3$ is coupled with a coherent CW at $\lambda 3$+60GHz generated by the remote 60GHz oscillator (OSC) wavelength feeder [31]. The phase-formatted XOR signal is sent via SMFs of 4 km to the RAU receiver (Rx) where it was converted to RF data by utilizing the beating at the PD [31]. The output is filtered by a BPF centered at 60 GHz and transmitted through an assumed wireless link to the user's Rx, where the electrical decoding is performed.

Figure 7(a) presents the user's Tx which produces the DQPSK-SCM data signal. Each user's Tx comprises a PPG loaded with a 4 Gb/s NRZ 2^7–1 PRBS, so as to form the electrical data. A serial-to-parallel distributor is fed with the output stream of the PPG and synchronously splits it into the two output streams (u and v), each having a data rate of 2Gb/s, resulting in a total bit rate equal to 4 Gb/s. The DQPSK differential encoding unit converts the u- and v-constituents into the respective I and Q signals, which then are inserted into the electrical phase modulators (PMs) to modulate the 60 GHz signal coming from a local oscillator (LO). RF signals coming from the LO have a relative phase difference of 90°, so as the DQPSK-RF signal to be generated after the combination of the phase-formatted I and Q streams.

Figure 7(b) depicts the user's Rx that is responsible for the down-conversion and the decoding of the incoming NC-encoded stream. Each user's Rx receives the phase-encoded XOR signal from the RAU and splits it into two identical signals. Those signals are multiplied with the respective in-phase signals originating from the LO and having a relative phase difference of 90°. The phase matching of the 60 GHz signals coming from the LO and the splitter is achieved by the phase shifters (PSs). The power regulation of the down-converted OOK u- and v-constituents, exiting the multiplier, is provided by a DC source. The resultant signals are filtered by low-pass filters (LPFs) and inserted in the electrical XOR gates where the decoding process is performed by the XOR operation between the NC encoding signals and a local copy of user's data. In this way, user A extracts the data B

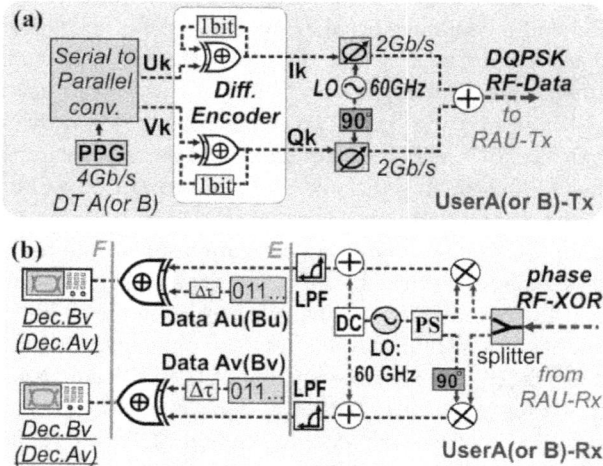

Figure 7. (a) Setup of the users' transmitter (Tx) generating the DQPSK-RF data and (b) setup of the users' receiver (Rx) performing the decoding operation. Stages E and F are the stages where the down-converted XOR-u (XOR-v) and the final decoded data signals are generated, respectively.

constituents, while user B decodes the data A u and v streams. The synchronization of the NC encoding signals and the user's data is achieved by the use of a time delay ($\Delta\tau$).

The simulations were carried out with the VPI photonics software suite [32], using as input the response of a custom-made SOA-MZI model [33] that matches the experimental measured response of a 1600-µm long hybrid-integrated SOA. The input power levels that were used were −1 dBm for the control signals and −4 dBm for the probe light, for all OOK SOA-MZIs. Both SOAs of the OOK-XOR and OOK-$\overline{\text{XOR}}$ MZI gates were driven by current values of 250 mA and had a recovery time of 100 ps, significantly longer than the 16.67 ps period of the 60 GHz SC. The SOAs of the phase regeneration XOR gates are driven by a 300 mA DC current and had an 80 ps recovery time.

4.2. Simulated results for synchronous operation

Figure 8 shows the time traces, eye diagrams and spectra obtained at various stages of the network coding-based 2 Gbaud RoF link during synchronous operation. The indicative patterns used for the simulation results are "1011111001" and "1010100110" for the u- and v- components of data A, while "1011010110" and "0100001100" were used for the data Bu

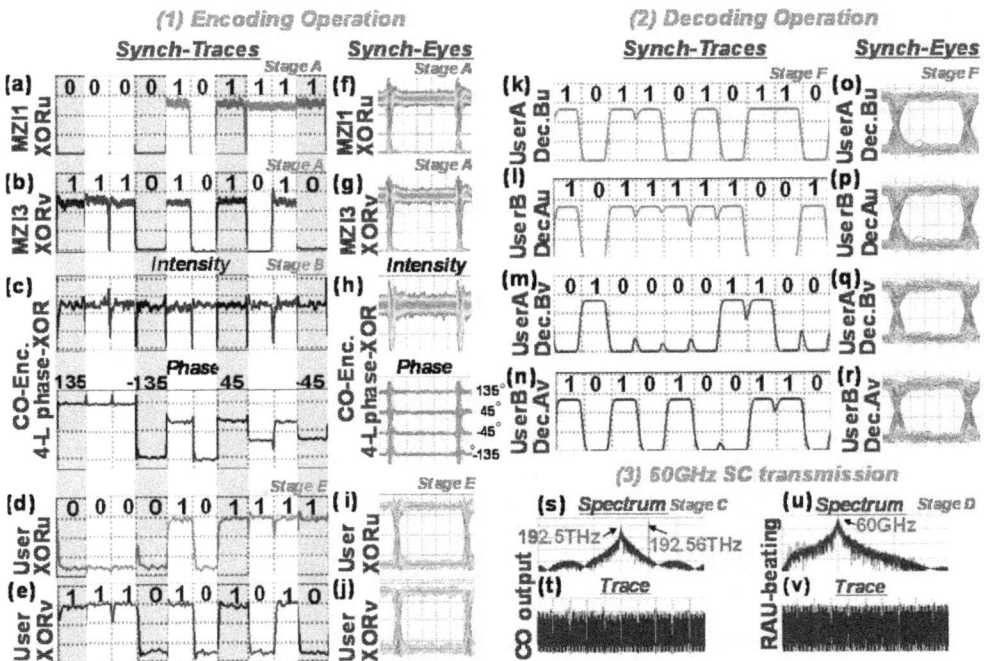

Figure 8. Results for synchronous AOPNC operation: (a)–(e) time traces of the encoding operation, (f)–(j) respective eye diagrams of encoding operation, (k)–(n) time traces of the decoding operation, (o)–(r) respective eye diagrams of decoding operation, (s) spectrum and (t) time trace of the SCM encoding signal at the CO's output and (u) spectrum and (v) time trace of the SCM RF encoding signal at the output of the RAU's Rx. 500 ps/div (traces) and 100 ps/div (eye diagrams).

and Bv components, respectively. **Figure 8(a)** and **(b)** shows the OOK-XORu and OOK-XORv time traces exiting the SOA-MZI-1 and SOA-MZI-3, respectively. Pulses are observed in the XOR streams, when data A and B constituents have different logical bit representations, while the power level is equal to 0 when data A and B exhibit the same logical value. The intensity and phase traces of the NC-phase-formatted stream at the output of the regenerator are shown in **Figure 8(c)**. The intensity envelope reveals a constant "high" power level, with small duration sub-bit dips generated by the transition of the differential phase between the SOA-MZI-5 and SOA-MZI-6 branches from +90° to −90°. The phase time trace presents the NC phase-encoded resultant signal, whose optical phase ϕ can take one of the four values: [−135, 135, −45 and 45°], corresponding to the logical bit pairs: "XORu, XORv"= ["00", "01", "10" and "11"], respectively. The grey markers highlight the encoding scheme, where the absence of the OOK-XORu and −XORv pulses is imprinted as −135°, the XORv pulses as 135°, the XORu pulses as −45° and the existence of both XORu, XORv pulses as 45°. **Figure 8(d)** and **(e)** depicts the electrical OOK down-converted XORu and XORv streams at the output of the low-pass filter (LPF) in the end-user receiver. These traces confirm the successful conversion of the four-level phase-formatted XOR signal to two binary NRZ-OOK XORu and XORv streams.

Figure 8(f) and **(g)** illustrates the eye diagrams of the OOK-XORu and XORv components exiting the SOA-MZI-1 and SOA-MZI-3. The eye diagrams exhibit an ER equal to 11.7 dB, an AM of 1.1 dB, a pulse overshoot (PO) of 1.4dB and a jitter of 23ps, showing a limited impact of pattern effects, stemming from the signal processing by SOAs. **Figure 8(h)** depicts the intensity eye diagram and the phase eye diagram of the NC-encoded phase-formatted signal exiting the phase regenerator. The intensity eye diagram exhibits an AM equal to 1.8 dB and a PO of 2.1 dB, while the respective phase eye diagram shows the four different phase levels of the encoded signal, revealing a relatively small-phase fluctuation of 1.5° from the expected phase values. In both diagrams, small duration dips at the beginning of the symbol pulse are observed, without yielding, however, any data loss. **Figure 8(i)** and **(j)** shows the eye diagrams of the electrical down-converted OOK XORu and XORv signal streams at the output of the LPF in the end-user Rx. The binary OOK XOR signals reveals an ER, AM, PO and jitter equal to 9 dB, 0.7 dB, 0.9 dB and 32 ps, respectively.

Figure 8(k) and **(l)** illustrates the decoded data Bu and Au streams, while **Figure 8(m)** and **(n)** depicts the complementary decoded data Bv and Av traces, each generated by the bit-wise XOR between the NC-encoded XOR trace and the local copy of the user's data pattern. **Figure 8(o)** and **(p)** shows the eye diagrams of the decoded data Bu or Au signals, reporting an open eye with an ER, AM and jitter equal to 8.7 dB, 1.1 dB and 40 ps, respectively, with the dips appearing at the beginning of the pulses. Similarly, **Figure 8(q)** and **(r)** depicts the eye diagrams of the decoded data Av and Bv, respectively, both exhibiting an ER of 8.7 dB, an AM of 1.1 dB and a jitter of 40 ps.

Figure 8(s) represents the optical spectrum of the phase-encoded signal at the output of the CO after the coupling of the four-level XOR signal with a continuous wave coming from the 60 GHz OSC wavelength feeder. The phase-formatted signal has a center wavelength of $\lambda3$=1557.36nm (192.5 THz), while the continuous wave is emitted at the wavelength of 1556.88 nm (192.560 THz), resulting in a frequency spacing equal to 60 GHz. The time trace at the same stage is illustrated in **Figure 8(t)**, showing an envelope with a constant power level, modulated by the sub-carrier of 60 GHz. **Figure 8(u)** shows the electrical spectrum of the beating signal at 60 GHz, produced by the photodiode and sent to the receiver of the end-user, while **Figure 8(v)** illustrates the respective time trace, showing the oscillations of the 60 GHz electrical signal below the constant power envelope.

4.3. Simulated results for asynchronous operation

In this section, the asynchronous operation was evaluated for different sub-bit time offsets between the two data signals. **Figure 9** includes the time traces and eye diagrams of the asynchronous encoding and decoding operation for a time offset of 0.5 of the symbol time duration (250ps). **Figure 9(a)** shows the OOK-XORu stream, generated after the XOR operation between the data Au and the delayed by 250 ps data Bu constituents at the output of the SOA-MZI-1, while **Figure 9(b)** illustrates the OOK-XORv signal exiting the SOA-MZI-3. As it is highlighted by the grey markers, the data asynchrony generates sub-bit pulses and dips at the encoded streams, which are said to be "interrupted". **Figure 9(c)** illustrates the intensity and phase traces of the four-level phase-formatted XOR signal exiting the regeneration stage after the recombination of the binary phase-XORu and phase-XORv signals. The intensity trace exhibits a constant power envelope with small duration sub-bit dips generated from the

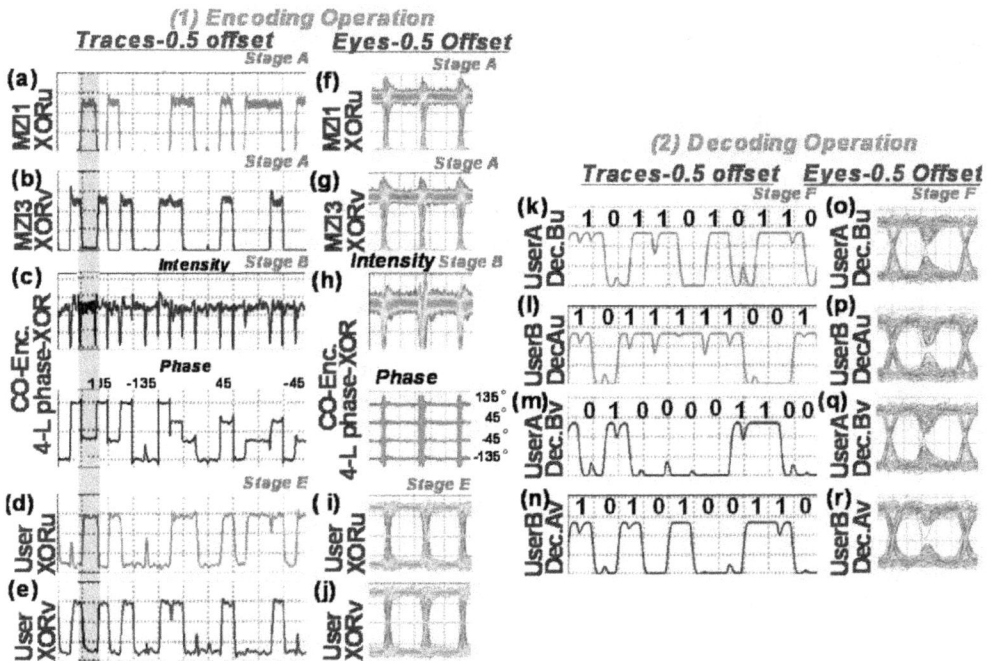

Figure 9. Results for the asynchronous AOPNC operation with a temporal offset equal to $0.5\tau_{bit}$: (a)–(e) time traces of the encoding operation, (f)–(j) respective eye diagrams of encoding operation, (k)–(n) time traces of the decoding operation, (o)–(r) respective eye diagrams of decoding operation. 500 ps/div (traces) and 100 ps/div (eyes).

transitions of the differential phase from +90° to −90° between the SOA-MZI-5 and/or SOA-MZI-6 branches. The phase trace shows the four different phase levels of the encoded signal, exhibiting sub-bit phase pulses and dips with a duration equal to 250 ps. **Figure 9(d)** and **(e)** depicts the XOR-u and XOR-v signals that were at the same time OOK- and down-converted by multiplying the received XOR streams with the respective in-phase 60 GHz signals generated by the LO at stage E.

Figure 9(f)–(j) shows the eye diagrams of the asynchronous encoding process. Particularly, **Figure 9(f)** and **(g)** illustrates the eye diagrams of OOK-XOR-u and XOR-v streams for a temporal offset equal to 0.5, exhibiting an ER of 11.7 dB, an AM of 1.1 dB, a PO of 1.4 dB and a jitter of 25 ps. As shown, an intersection is observed at both eye diagrams after a time delay equal to 250 ps from the beginning of the pulse. This intersection that has a jitter equal to 21 ps is formed by both sub-bit pulse falls and risings during the asynchronous XOR operation. **Figure 9(h)** illustrates the intensity and phase eye diagrams of the NC-phase-encoding signal at the output of the regenerator. An intensity envelope with an AM of 1.8 dB and PO of 2.1 dB is shown, exhibiting sub-bit dips when the relative phase of SOA-MZI-5 and/or SOA-MZI-6 is changed from 90 to −90° and vice versa. The respective phase eye diagram shows the four different phase levels of the encoded signal, revealing a small phase fluctuation equal to 1.5°. The electrical eye diagrams of the down-converted OOK-XORu and XORv signals at the users' Rx are depicted in **Figure 9(i)** and **(j)**, revealing an ER, AM, PO and a jitter equal to 9, 0.7, 0.9 dB and 32 ps, respectively.

Figure 9(k)–(r) shows the time traces and eye diagrams of the asynchronous decoding operation. **Figure 9(k)** and **(l)** depicts the decoded data Bu and Au components after the digital XOR operation between the encoded pattern of XORu and the initial pattern either of data Au or data Bu at the users A and B Rx, respectively. Similarly, **Figure 9(m)** and **(n)** illustrates the decoded data Bv and Av at the user A and B receivers, respectively. In that operation, it is shown that despite the interruptions which appear as "parasitic" pulses or dips at the NC-encoded signals, both the components of data A and B were at the end correctly recovered with the decoded data B components having a delay equal to the time offset. **Figure 9(o)** and **(p)** depicts the eye diagrams of the decoded data Bu and Au, respectively, both exhibiting an ER of 8.7 dB, an AM of 1.1 dB and jitter of 40 ps, with short duration dips and spikes appearing after time delay equal to 250 ps from the beginning of the symbol. Similarly, **Figure 9(q)** and **(r)** shows the eye diagrams of the decoded data Bv and Av, reporting similar eye characteristics (ER=8.7 dB and AM=1.1 dB, jitter=40 ps).

The asynchronous decoding operation was evaluated by carrying out BER measurements. **Figure 10(a)** shows the BER measurements versus the received RF power, for the decoded Au and Bu streams, when the time offset between the two data is equal to 0.25, 0.5 and 0.75 of the symbol duration. All BER curves show error-free operations at 10^{-9}. The BER curves reveal similar performance between synchronous and asynchronous operation, owing to the dips and spikes that were present at the edge of the pulse during synchronous operation being shifted within the duration of the pulses, however, without affecting the other pulse characteristics, such as ER, AM, PO, jitter and noise. Similarly, **Figure 10(b)** depicts the BER

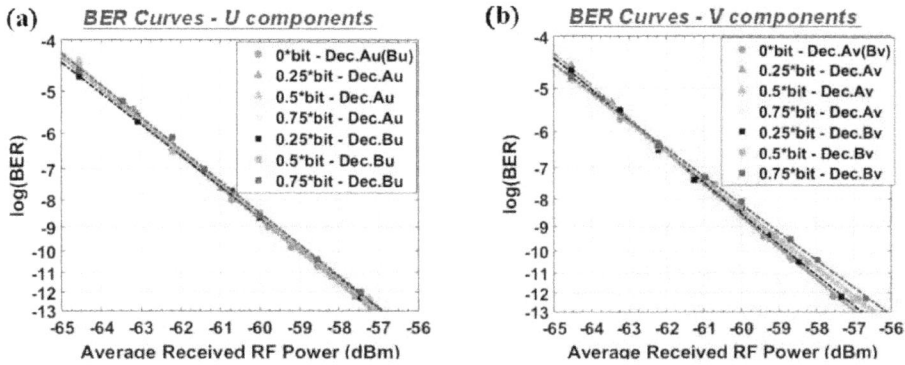

Figure 10. BER curves of the: (a) decoded data au and Bu and (b) decoded data Av and Bv components for various time offsets of 0.25bit, 0.5bit and 0.75bit in asynchronous operation.

curves of the decoded Av and Bv streams for time offsets equal to $0\tau_{symbol}$, $0.25\tau_{symbol}$, $0.5\tau_{symbol}$ and $0.75\tau_{symbol}$, showing error-free operations with negligible power penalty between the different BER curves. It is evident that the performance of the proposed AOPNC scheme remains similar even in the case of asynchronous packets reaching the network coder.

5. Conclusion

In this chapter, the concept of digital all-optical physical-layer network coding (AOPNC) was presented, targeting the future high-throughput radio-over-fiber (RoF) networks. In this scheme, the bitwise network coding (NC) is performed on the fly at the central office (CO) and the resultant packet is broadcasted at the wireless users, where the decoding takes place. The applicability of the AOPNC scheme between OOK-sub-carrier-modulated (SCM) data signals was confirmed by an experimental demonstration, employing a 10 GHz SC and an all-optical XOR gate as the digital NC encoder. An AOPNC scheme capable of performing the digital encoding and decoding between DQPSK-SCM data signals was also described. In this scheme the scenario of all optical encoding for 60 GHz SC used in mm-wave communications, followed by electrical decoding at the end users, was evaluated via physical-layer simulations. It should be noted that the described all-optical network Coding concept may in principle be applied also in RoF systems using DPSK-SCM and dual polarization (DP)-DQPSK-SCM modulation formats.

Acknowledgements

This work has been supported by the European FP7-PEOPLE-2013-IAPP project COMANDER (contract no. 612257) and the H2020-ITN-2016 project 5 G-STEP-FWD (contract no. 722429).

Author details

Charikleia Mitsolidou[1,2]*, Chris Vagionas[1,2], Dimitris Tsiokos[1,2], Nikos Pleros[1,2] and Amalia Miliou[1,2]

*Address all correspondence to: cvmitsol@csd.auth.gr

1 Department of Informatics, Aristotle University of Thessaloniki, Thessaloniki, Greece

2 Center for Interdisciplinary Research and Innovation, Aristotle University of Thessaloniki, Thessaloniki, Greece

References

[1] Cisco. Cisco Visual Networking Index: Global Mobile Data Traffic Forecast Update. 2016-2021 [Internet]. 2017. Available from: http://www.cisco.com/c/en/us/solutions/collateral/service-provider/visual-networking-index-vni/mobile-white-paper-c11-520862. html [Accessed: Jan 17, 2018]

[2] Ericsson. State and Future of the Mobile Networks [Internet]. 2017. Available from: https://www.ericsson.com/mobility-report/state-and-future-of-the-mobile-networks [Accessed: Jan 17, 2018]

[3] Ghazisaidi N, Maier M. Fiber-wireless (FiWi) access networks: Challenges and opportunities. IEEE Network. 2011;**25**:36-42. DOI: 10.1109/MNET.2011.5687951

[4] Novak D, Waterhouse RB, Nirmalathas A, Lim C, Gamage PA, Clark TR, Dennis ML, Nanzer JA. Radio-over-fiber Technologies for Emerging Wireless Systems. Journal of Quantum Electronics. 2015;**52**. DOI: 10.1109/JQE.2015.2504107

[5] Xu K, Wang R, Dai Y, Yin F, Li J, Ji Y, Lin J. Microwave photonics: Radio-over-fiber links, systems, applications. Photonics Research. 2014;**2**:B54-B63. DOI: 10.1364/PRJ.2.000B54

[6] Caballero A, Zibar D, Sambaraju R, Guerrero Gonzalez N, Tafur Monroy I. Engineering rules for optical generation and detection of high speed wireless millimeter-wave band signals. In: Proceedings of the European Conference on Optical Communications (ECOC '11); Geneva, Switzerland: IEEE; 18-22 September 2011

[7] Pleros N, Vyrsokinos K, Tsagkaris K, Tselikas ND. A 60 GHz radio-over-fiber network architecture for seamless communication with high mobility. IEEE/OSA Journal of Lightwave Technology. 2009;**27**:1957-1967. DOI: 10.1109/JLT.2009.2022505

[8] Chen L-K, Li M, Liew SC. Breakthroughs in photonics 2014: Optical physical-layer network coding, recent developments, and challenges. IEEE Photonics Journal. 2015;**7**. DOI: 10.1109/JPHOT.2015.2418264

[9] Liu ZX, Lu L, You L, Chan CK, Liew SC. Optical physical-layer network coding over fiber-wireless. In: Proceedings of the European Conference on Optical Communications (ECOC '13); London, UK: IEEE; 22-26 September 2013

[10] Manley ED, Deogun JS, Xu L, Alexander DR. All-optical network coding. IEEE/OSA Journal of Optical Communications and Networking. 2010;**16**:175-191. DOI: 10.1364/JOCN.2.000175

[11] Katti MS, Rahul H, Hu W, Katabi D, Médard M, Crowcroft J. Xors in the air: Practical wireless network coding. IEEE/ACM Transactions on Network. 2008;**16**:497-510. DOI: 10.1145/1159913.1159942

[12] Miller K, Biermann H, Woesner H, Karl H. Network coding in passive optical networks. In: Proceedings of the IEEE International Symposium on Network Coding (NetCod '10); Toronto, Canada: IEEE; 9-11 June 2010

[13] Fouli K, Maier M, Médard M. Network coding in next-generation passive optical networks. IEEE Communications Magazine. 2011;**49**:38-46. DOI: 10.1109/MCOM.2011.6011732

[14] Belzner M, Haunstein H. Network coding in passive optical networks. In: Proceedings of the European Conference on Optical Communications (ECOC '09); Vienna, Austria: IEEE; 20-24 September 2009

[15] Thinniyam RS, Kim M, Médard M, O'Reilly U-M. Network coding in optical networks with O/E/O based wavelength conversion. In: Proceedings of the Optical Fiber Communication Conference and Exposition/National Fiber Optic Engineers Conference (OFC/NFOEC '10); San Diego, CA, USA: OSA/ IEEE; 21-25 March 2010

[16] Qu Z, Ji Y, Bai L, Sun Y, Fu J. Key module for a novel all-optical network coding scheme. Chinese Optics Letters. 2010;**8**:753-756. DOI: 10.3788/COL20100808.0753

[17] Hisano D, Maruta A, Kitayama K. Demonstration of all optical network coding by using SOA MZI based XOR gates. In: Proceedings of the Optical Fiber Communication Conference and Exposition/National Fiber Optic Engineers Conference (OFC/NFOEC'13); Anaheim, CA, USA: OSA/ IEEE; 17-21 March 2013

[18] An Y, Ross FD, Peucheret C. All-optical network coding for DPSK signals. In: Proceedings of the Optical Fiber Communication Conference and Exposition/National Fiber Optic Engineers Conference (OFC/NFOEC '13); Anaheim, CA, USA: OSA/ IEEE; 17-21 March 2013

[19] Lu G-W, Hongxiang JQ, Gazi WYJ, Sharif M, Yamaguchi S. Flexible and re-configurable optical three-input XOR logic gate of phase-modulated signals with multicast functionality for potential application in optical physical-layer network coding. Optics Express. 2016;**24**:2299-2306. DOI: 10.1364/OE.24.002299

[20] Liu ZX, Li M, Lu L, Chan C-K, Liew SC, Chen L-K. Optical physical-layer network coding. IEEE Photonics Technology Letters. 2012;**24**:1424-1427. DOI: 10.1109/LPT.2012.2204972

[21] Mitsolidou C, Vagionas C, Ramantas K, Tsiokos D, Miliou A, Pleros N. Digital optical physical-layer network coding for mm-wave radio-over-fiber signals in fiber-wireless networks. IEEE/ OSA Journal of Lightwave TechnologyJ. Lightwave Technol. 2016;**34**:4765-4771. DOI: 10.1109/JLT.2016.2585673

[22] Caballero A, Zibar D, Monroy IT. Performance evaluation of digital coherent receivers for phase modulated radio-over-fiber links. IEEE/ OSA Journal of Lightwave Technology. 2011;**29**:3282-3292. DOI: 10.1109/JLT.2011.2167595

[23]   Sambaraju R, Zibar D, Caballero A, Monroy IT, Alemany R, Herrera J. 100-GHz wireless-over-fibre links with up to 16 Gb/s QPSK modulation using optical heterodyne generation and digital coherent detection. IEEE Photonics Technology Letters. 2010;**2010**:1650-1652. DOI: 10.1109/LPT.2010.2076801

[24]   Mitsolidou C, Pleros N, Miliou A. All-optical digital physical-layer network coding for DPSK mm-wave radio-over-fiber networks. In: Proceedings of the 19th International Conference on Transparent Optical Networks (ICTON '17); Girona, Spain. 2-6 July 2017

[25]   Mitsolidou C, Pleros N, Miliou A. Digital all-optical physical-layer network coding for 2Gbaud DQPSK signals in mm-wave radio-over-fiber networks. Optical Switching and Networking. 2017; (In press) ISSN: 1573-4277. DOI: 10.1016/j.osn.2017.10.002

[26]   Spyropoulou M, Pleros N, Miliou A. SOA-MZI-based non-linear optical signal processing: A frequency domain transfer function for wavelength conversion, clock recovery and packet envelope detection. IEEE Journal of Quantum Electronics. 2011;**47**:40-49. DOI: 10.1109/JQE.2010.2071411

[27]   Chang K, Liu C. 1-100 GHz microwave photonics link technologies for next-generation WiFi and 5G wireless communications. In: Proceedings of the International Topical Meeting on Microwave Photonics (MWP '13); Alexandria, VA, USA: IEEE; 28-31 October 2013

[28]   Vorreau P, Marculescu A, Wang J, Bottger G, Sartorius B, Bornholdt C, Slovak J, Schlak M, Schmidt C, Tsadka S, Freude W, Leuthold J. Cascadability and regenerative properties of SOA all-optical DPSK wavelength converters. IEEE Photonics Technology Letters. 2006;**18**:1970-1972. DOI: 10.1109/LPT.2006.880714

[29]   Wang G, Yang X, Hu W. All-optical logic gates for 40Gb/s NRZ signals using complementary data in SOA-MZIs. Optics Commununications. 2013;**290**:28-32. DOI: 10.1016/j.optcom.2012.10.047

[30]   Dennis ML, Nanzer JA, Callahan PT, Gross MC, Clark TR, Novak D, Waterhouse RB. Photonic upconversion of 60 GHz IEEE 802.15.3c standard compliant data signals using a dual-wavelength laser. In: Proceedings of the Annual Meeting of IEEE Photonics Society; Denver, CO, USA. 7-11 November 2010

[31]   Seagraves E, Berry C, Qian F. Robust mobile WiMax preamble detection. In: Proceedings of the Military Communications Conference (MILCOM '08); San Diego, CA, USA: IEEE; 16-19 November 2008

[32]   VPIphotonics. VPIphotonics Official Wedsite [Internet]. 2018. Available from: http://www.vpiphotonics.com/index.php [Accessed: Jan 17, 2018]

[33]   Vagionas C, Fitsios D, Vyrsokinos K, Kanellos GT, Miliou A, Pleros N. XPM- and XGM-based optical RAM memories: Frequency and time domain theoretical analysis. IEEE Journal of Quantum Electronics. 2014;**47**:683-697. DOI: 10.1109/JQE.2014.2330068

Chapter 3

Network Coding for Distributed Antenna Systems

Rafay Iqbal Ansari, Muhamad Arslan Aslam,
Syed Ali Hassan and Chrysostomos Chrysostomou

Additional information is available at the end of the chapter

http://dx.doi.org/10.5772/intechopen.76323

Abstract

The mushroom growth of devices that require connectivity has led to an increase in the demand for spectrum resources as well as high data rates. 5G has introduced numerous solutions to counter both problems, which are inherently interconnected. Distributed antenna systems (DASs) help in expanding the coverage area of the network by reducing the distance between radio access unit (RAU) and the user equipment. DASs that use multiple-input multiple-output (MIMO) technology allow devices to operate using multiple antennas, which lead to spectrum efficiency. Recently, the concept of virtual MIMO (VMIMO) has gained popularity. VMIMO allows single antenna nodes to cooperate and form a cluster resulting in a transmission flow that corresponds to MIMO technology. In this chapter, we discuss MIMO-assisted DAS and its utility in forming a cooperative network between devices in proximity to enhance spectral efficiency. We further amalgamate VMIMO-assisted DAS and network coding (NC) to quantify end-to-end transmission success. NC is deemed to be particularly helpful in energy constrained environments, where the devices are powered by battery. We conclude by highlighting the utility of NC-based DAS for several applications that involve single antenna empowered sensors or devices.

Keywords: network coding, D2D, opportunistic networks, cooperative communication, energy-efficiency

1. Introduction

The evolution of 5G networks will open up numerous new opportunities in terms of applications that require higher data rates, reliability and low latency. Spectrum limitation is one of the obstacles that could impede the growth of 5G networks. Several solutions have been proposed to overcome the spectrum scarcity. The efficient utilization of available spectrum resources has gained attention of the research community. Distributed antenna systems (DAS) are considered as one of the solutions for ensuring efficient utilization of spectral resources [1] and providing high data rates. DAS are based on a dense deployment of remote access units (RAUs) in a cellular network, thereby reducing the distance between the users and the RAUs. The RAUs are connected to a central control module through high rate dedicated links. The presence of the users in a close vicinity allows the RAUs to transmit at low power, which leads to energy savings. One of the underlining features of 5G networks is the concept of *green communications* [2]. DAS can help in realizing green communications by ensuring energy-efficient network operation. Several works can be found in literature that addresses the energy-efficiency (EE) of DAS [3, 4]. Another key technology is the multiple-input multiple-output (MIMO), which signifies the presence of multiple antennas in the form of an antenna array [5]. MIMO-assisted DAS provides a robust solution for network connectivity by mitigating the impact of interference and allowing the transmission of multiple data streams simultaneously.

Recently, the concept of virtual MIMO (VMIMO) has been explored for dense network environments. In VMIMO, the single antenna sensors or user equipments cooperate to form a cluster. The cluster is then considered as a single MIMO system that transmits packets to the adjacent cluster in a multi-hop manner [6]. In this chapter, we consider the VMIMO-assisted DAS and employ network coding (NC) to gauge the network performance. NC techniques can further improve the performance of the network with regards to EE and spectral-efficiency (SE) by avoiding packet retransmission. Packets sent by the source nodes can be combined to form coded messages, which are then sent to the destination. NC is particularly helpful in energy constrained environments, where the devices possess limited battery power. NC can be employed in DAS to enhance the network throughput and improve the end-to-end success of the network [7]. In DAS involving multi-hop transmissions, NC can minimize the transmission delay by allowing cooperation between nodes. In this chapter, we consider two DAS environments (1) Low-density DAS, and (2) High-density DAS. Applications for low-density DAS include device-to-device (D2D) communications, where a few antenna elements in the form of user equipments (UEs) are trying to connect to each other. On the other hand, a typical wireless sensor network, where a large number of nodes are deployed in an area, provides an example of high-density DAS. The remainder of the chapter is organized as follows. First, we enlist the benefits of DAS and applications that utilize DAS to enhance coverage and quality-of-service (QoS). Next, we discuss MIMO-assisted DAS and the concept of VMIMO, which is followed by the evaluation of NC in the aforementioned DAS environments. The results quantify the end-to-end transmission success and probability distribution at different network parameters, providing a designer's perspective for VMIMO-assisted DAS.

2. Applications, benefits and limitations of DAS

DAS can help in realizing several new applications. There are several benefits associated with DAS; however, there are also some limitations that need to be taken into consideration while designing networks based on DAS. Multi-service indoor DAS (MS-IDAS) is a class of DAS, which is particularly helpful in applications that involve indoor environments, such as shopping malls, restaurants and bus stations. The hardware modules for such environments are manufactured by keeping in view the esthetics of the environment [8]. DAS are helpful in network applications that involve mobility. Similarly, DAS allow the expansion of coverage to areas, which cannot be covered by the traditional network due to the blockage by physical structures. The geographical areas that undergo blockages are referred to as *coverage holes*. DAS can alleviate the situation with regards to coverage holes and can help in realizing the concept of ubiquitous connectivity. For example, the network connectivity can be ensured on a high speed train by employing the concept of DAS [9].

Massively DAS (MDAS) lead to higher network diversity but at the cost of higher computational complexity. A coordinated antenna selection (CAS) is required to control the operation of a MDAS [10] by selectively activating the antenna elements that provide the best link quality. CAS can help in mitigating the impact of co-channel interference by coordinating the antenna activation; however, one of the limitations of designing a CAS for MDAS is the perfect channel state information (CSI) that is required for scheduling the transmissions. The practical limitations with regards to acquiring perfect CSI restrict the MDAS operation by allowing it to serve less number of users. Channel estimation or prediction techniques are required for smooth operation of a MDAS.

The geographical distribution of RAUs allows spatial degrees of freedom but at the cost of higher computational complexity. In a multi-user DAS-based network, the users with the best channel conditions are served while the users suffering from adverse channel conditions are not scheduled, leading to unfairness in resource allocation [11]. The concept of ubiquitous connectivity for 5G networks is also compromised due to unfair resource allocation. A fair scheduling mechanism is necessary to ensure that the QoS requirements of all the users are met with efficient utilization of resources. Moreover, in an internet-of-things (IoT) environment it is imperative that all the devices and users are able to connect to the network. Concluding, the benefits of DAS include

- spatial diversity,

- efficient spectrum utilization,

- energy-efficiency,

- higher network rate,

- enhanced end-to-end success,

- ubiquitous connectivity,

- interference mitigation.

DAS can go a long way in realizing new applications that would arise with the introduction of 5G networks. An illustration of a DAS based network is shown in **Figure 1**, where the DAS nodes are connected through fiber links to the central processing unit. DAS allows to extend the coverage to geographical areas which are not covered by the traditional networks due to physical blockages.

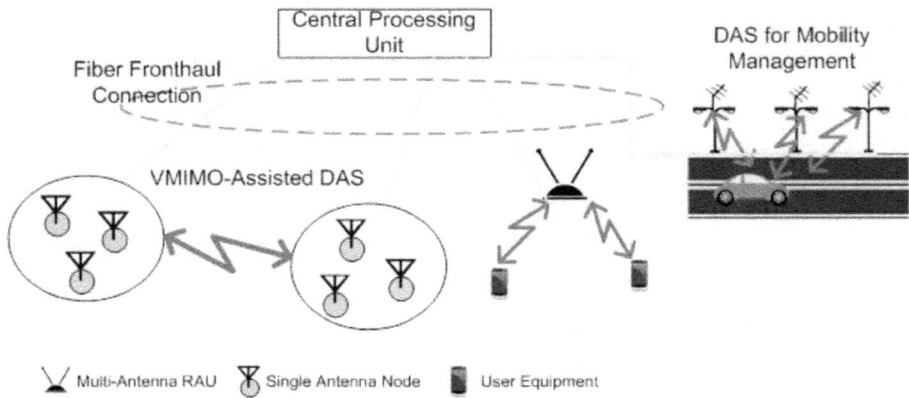

Figure 1. Example of DAS-based network.

3. MIMO-assisted DAS

Based on the number of antenna elements, the DAS can be further categorized as

- DAS based on single-antenna RAU
- DAS based on multiple-antenna RAU

Initially, the research related to DAS focused on single-antenna RAU [12]. However, the introduction of MIMO technologies opened up a new arena with regards to spectrum utilization. The integration of the concept of DAS empowered by MIMO envisioned significant gains in terms of network reliability. The presence of multiple antennas at the RAUs and users allows multiple links to be established between the user and the RAU, leading to higher rates. Moreover, the number of users also impact the performance of the MIMO assisted DAS. If there are multiple users within the coverage range of the RAU, beamforming is conducted to serve all the users. The network performance is impacted by the increase in the antenna elements at the RAU [13].

MIMO can be categorized into: (1) single-user MIMO (SU-MIMO), (2) multi-user MIMO (MU-MIMO). The impact of channel environment on the network performance is less pronounced in case of MU-MIMO, as compared to SU-MIMO. The reason for such behavior lies in the multi-user diversity that can be achieved through MU-MIMO [14]. In SU-MIMO, the resources are dedicated to a single user for achieving higher capacity. Spatial multiplexing and beamforming aid in forming high capacity transmission links. Spatial multiplexing allows the transmission of multiple streams, where these streams undergo spatial processing at the receiver [15]. On the other hand, MU-MIMO allows allocation of resources to multiple users leading to multi-user diversity and performance gains as compared to SU-MIMO. MU-MIMO is particularly helpful in scenarios marked by high traffic. MU-MIMO leads to increased throughput, increased diversity gain and reduced costs as compared to SU-MIMO.

MIMO involves an integration of multiple antenna elements at access points (APs) and base stations (BSs), leading to higher network capacity. Transmission beams with high directivity are formed through beamforming, leading to low interference and high transmission gain communication links. The concept of MIMO has been included in several wireless network standards such as IEEE 802.11n, 802.11 ac WLAN, 802.16e (Mobile WiMAX), 802.16 m (worldwide interoperability for microwave access (WiMAX), 802.20 mobile broadband wireless access (MBWA), 802.22 (WRAN), 3GPP long-term evolution (LTE) and LTE-Advanced evolved universal terrestrial radio access(E-UTRA) [16]. Resource allocation techniques for massive MIMO have been devised to ensure efficient utilization of resources. The resource allocation is conducted by keeping in view the requirements of the desired QoS of the individual users. Moreover, the MIMO is backed by spatial diversity and multiplexing techniques to improve the network performance. Spatial multiplexing is required in multi-user MIMO to enjoy the gains of spatial diversity. In multi-user MIMO, simultaneous data streams are sent to multiple users to increase the network capacity. Below, we list some challenges related to massive MIMO:

- manufacturing low cost base stations

- ensuring hardware compatibility

- designing lower antenna size

- acquiring CSI

- designing low power base stations

Third generation partnership project (3GPP) has defined the key features of MIMO and refers to two-dimensional (2D) antenna array structures as full dimensional MIMO (FD-MIMO). FD-MIMO involves 3D channel propagation, where the path loss is dependent upon the height and distance of the user from the AP. The elevation angle is also one of the aspects that is included in the 3D channel model. An increase in the number of antennas allows a simple interference management by using a precoder [17]. The presence of a dense antenna array also allows network robustness in case of hardware failures [18]. SE and EE can be achieved by utilizing antenna arrays at the user and the BS, but the size of antenna array depends on the hardware compatibility and size of the device.

3.1. Virtual MIMO for multi-hop networks

5G networks would involve densification of devices and D2D networks are considered as one of the technologies that could alleviate the burden on the BS. The BS can be equipped with multiple-antennas but the D2D UE is limited to having a single-antenna capability due to size limitations. However, in the up-link transmission mode, multiple users can coordinate with each other to use the same sub-channels and create a virtual antenna array, leading to the concept of VMIMO [6]. The concept of VMIMO is also helpful in multi-hop networks involving cluster environments, e.g., D2D cluster networks [19]. Each cluster contains multiple nodes having single antennas. The cluster acts as a multi-antenna node and helps in transmitting the information cooperatively to the adjacent clusters [20]. The relay mechanisms that can be utilized to realize multi-hop transmission between clusters or nodes include amplify-and-forward (AF), compress-and-forward (CF) and decode-and-forward (DF). Another VMIMO architecture that has been proposed is the multi-hop relaying based on coding techniques. The relay node collects the information sent by the transmitters and forms a coded message that is shared with the destination in the next time slot [21].

4. Network coding for DAS

In Section 1.4, we built the case for DAS by presenting their advantages with regards to 5G networks and also discussed MIMO assisted DAS. In this section, we signify the utility of employing NC in DAS. We divide the study into two environments:

1. Low density DAS (D2D multi-hop networks)

2. High density DAS [opportunistic large array (OLA) multi-hop networks]

In the proceeding, we first provide a brief overview of the NC techniques that have been discussed in literature. Next, we describe the system models of both environments and quantify the performance gains achieved through NC in terms of end-to-end transmission success.

4.1. Network coding in low density DAS (D2D multi-hop networks)

D2D networks are based on a peer-to-peer (P2P) network between devices instead of relying on the BS for data transmission. The BS is responsible for supporting the control plane, while the devices establish a direct link to share the message with each other. Cooperative diversity could be exploited by D2D networks for ensuring reliable communication. NC-aided cooperative D2D networks enhance the success probability of end-to-end data delivery. In [22], NC is employed in BS-assisted D2D networks. **Figure 2** shows the system model, where the BS operates as a relay between the two D2D users. The D2D users transmit in the first two time slots. In the third time slot, the BS applies exclusive OR (XOR) to form a coded message and transmits it to the D2D users.

Generally, interference is considered as an impairment, but [23] introduces the concept of physical layer NC (PNC) in D2D networks, where interference is maneuvered positively to form coded messages. PNC scheme simply superimposes electromagnetic waves and forms a code. **Figure 3** highlights the PNC operation in a two-source, one-relay scenario. The first time slot is reserved for transmission of the messages x_1 and x_2 by the D2D users. The relay performs PNC to form a coded message x_3 in the second time slot. The conventional two-source relay networks utilize four time slots to complete information exchange. However, PNC-assisted two-source relay network realizes information exchange in two time slots, thereby enhancing the network capacity.

NC can be utilized to enhance performance in mobile cloud scenarios [24]. The authors highlight the diversity gains that can be achieved in high node density D2D networks. The results also show the benefits of D2D networks assisted by NC for providing live data transmission. The devices form a cooperative network and share chunks of data cooperatively to complete the downloading process. The proposed technique leads to energy savings and helps in avoiding delay in transmissions.

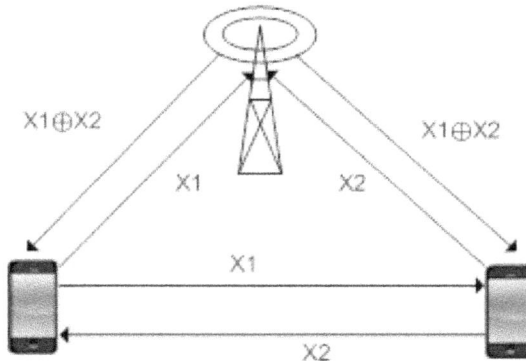

Figure 2. Network coding assisted D2D cooperative network [22].

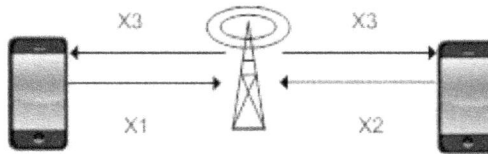

Figure 3. Physical layer network coding assisted D2D communication network [23].

In [25], the authors present the integration of caching techniques and NC-assisted D2D communication networks. Such networks are particularly helpful in realizing the proximity services. D2D users that require similar content can utilize such networks receive the desired content, cooperatively. Ref. [26] presents a comparison between D2D aided by space–time analog NC (STANC), traditional D2D networks and D2D aided by analog NC. The system model comprises of a relay that has two antennas and three D2D pairs. The relay employs amplify and forward technique to transmit the message to the destination. The desired information is recovered at the destination through zero forcing detection. The average sum rate is computed to highlight the utility of STANC as compared to other techniques.

4.1.1. Evaluation model

In our analysis, we consider a DAS with 3-D2D pair network aided by a relay node as shown in **Figure 4**. One transmission cycle comprises of four time slots. The source S_1 transmits in the first time slot. Sources S_2, S_3 and relay R transmit in the next three time slots, respectively [27]. The transmission is considered successful if the received SNR is greater than a threshold τ. The source S_1 broadcasts its message X_1 in the first time slot, which is received by all other receivers and sources, as shown in **Figure 4(a)**. In the next time slot, source S_2 formulates a coded message by employing NC, i.e., $a_1 X_1 + a_2 X_2$, and broadcasts the coded message as shown in **Figure 4(b)**. Linearly independent codewords are formed by choosing the coefficients a_1 and a_2 from a finite Galois field. In the same manner, S_3 forms a coded message $a_3 X_1 + a_4 X_2 + a_5 X_3$ and transmits in the third time slot (**Figure 4c**), followed by the transmission from relay R in the fourth time slot (**Figure 4d**). The destination receives multiple codewords and it can recover the intended information by utilizing Gaussian elimination technique.

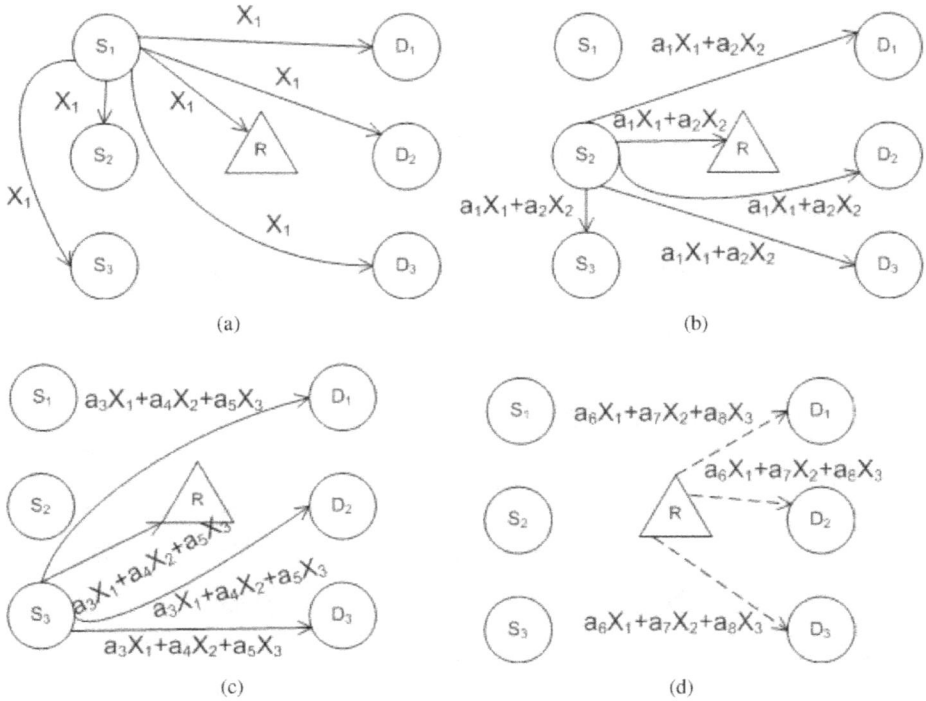

Figure 4. Relay-aided-network-coded (RANC) D2D network transmission model [27].

The transmission model presented in **Figure 4** is designed to provide diversity to the channels suffering from weakest links. In the case shown in **Figure 4**, the highest priority P_1 is assigned to D2D pair 1, as in the ideal case the destination D_1 can receive four codewords that contain the message X_1. Similarly, priorities P_2 and P_3 are assigned to the destinations D_2 and D_3, respectively. It is pertinent to mention that the transmission flow can be adapted dynamically according to the channel states. For example, if D2D pair 2 suffers from worse channel conditions, then priority P_1 could be assigned to it, leading to more diversity and hence an increase in success probability for the transmissions that are destined for D_2. Another criteria that drives the assignment of the priorities is the sensitivity of the information that is being shared over the network. If the information being transmitted is sensitive in nature, then a higher priority could be assigned to that D2D pair.

A comparison between traditional relay-aided D2D network and RANC is presented [27]. The aim of the comparison is to ascertain the deployment that best suits the QoS requirements in a particular network setting. **Figure 5** signifies the behavior of simple relay-aided D2D network and RANC versus the signal-to-noise ratio (SNR) margin. SNR margin is defined as SNR normalized by the threshold, τ. In these results, we assume that D2D pair 1 is assigned priority P_1. We determine the success probability for two different scenarios:

1. Network model excluding the impact of path loss

2. Network model including the impact of path loss

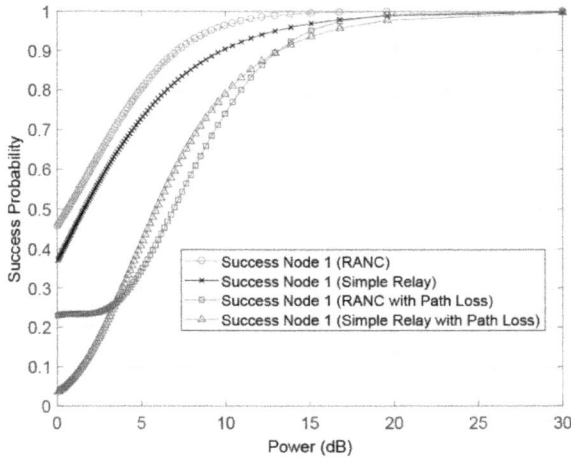

Figure 5. Success probability for D2D pair 1.

It can be observed that at a specific SNR margin, e.g., 15 dB, the performance of RANC and simple relay-aided D2D network can be analyzed. In this case, RANC outperforms the simple relay-aided D2D network for both aforementioned scenarios. These results could be particularly helpful in identifying the SNR margins that would be required for maintaining a particular end-to-end success probability in VMIMO-assisted DAS.

To further elaborate the findings, we present the performance comparison between simple relay-aided D2D and RANC D2D at different values of SNR margin. If $P^{D_i}_{s_{RANC}}$ and $P^{D_i}_{s_{relay-aided}}$ denote the success probabilities of RANC and simple relay-aided D2D network when calculated at a particular destination D_i, respectively, then the percentage improvement in success probability for RANC is given as

$$\text{percentage improvement} = \frac{P^{D_i}_{s_{RANC}} - P^{D_i}_{s_{relay-aided}}}{P^{D_i}_{s_{relay-aided}}} \times 100. \tag{1}$$

Table 1 highlights the comparative analysis of simple relay-aided D2D and RANC D2D. The results are presented at different SNR margins in the form of percentages calculated using (1.1). The negative entries in the table signify a performance degradation of RANC, while the positive entries signify the performance gains that can be achieved through RANC.

The performance of RANC at D2D pair 1 is analyzed for different fading characteristics. Note that λ characterizes the power of Rayleigh channel in a link. The fading characteristics are varied by changing the values of λ, where $\lambda = 0.2$ denotes the strongest gain. It can be observed that at higher values of SNR margin, RANC provides significant performance gains as compared to the simple relay-aided D2D network. Similar results are presented in **Table 2** for a scenario involving path loss. The results signify the utility of employing RANC when the links suffer from channel degradation. Dynamic priority assignment can help in increasing the end-to-end transmission success for links suffering from channel degradation.

SNR margin(dB)	0 (%)	10 (%)	15 (%)	20 (%)	25 (%)	30 (%)
D2D Pair 1	23.60	6.56	2.75	0.95	0.31	0.1
D2D Pair 2	−17.71	3.88	2.4	0.92	0.30	0.09
D2D Pair 3	−4.81	1.81	−4.23	0.06	0.21	0.09
D2D Pair 1, λ = 0.2	29.28	8.83	3.03	0.98	0.31	0.09
D2D Pair 1, λ = 0.4	28	8.2	2.9	0.90	0.31	0.09
D2D Pair 1, λ = 0.6	26.6	7.67	2.89	0.97	0.30	0.09
D2D Pair 1, λ = 0.8	25.18	7.11	2.82	0.96	0.30	0.09

Table 1. Analysis of simple relay-aided D2D with RANC D2D (excluding path loss) [27].

4.2. NC in high density DAS (OLA multi-hop networks)

Cooperative OLA networks allow a cluster of nodes to share the message with the adjacent cluster through multi-hop communication. OLA can be considered as a variant of DAS in network environments involving high density of nodes. In our analysis, we consider an OLA network which comprises of multiple source nodes that transmit the message to a common destination using a multi-hop topology [28]. We consider two network topologies for evaluation (1) Deterministic topology, and (2) Random topology.

Deterministic topology Consider a network topology shown in **Figure 6**, where two sources are deployed at a particular distance from each other. There are N relay nodes in a cluster (hop), where the average distance between each cluster is denoted by d. The number of clusters is denoted by n, while the destination is denoted by D. The source nodes operate at orthogonal frequencies to send the information to the first cluster of nodes. The relays in the first cluster employ DF mechanism to transmit the information to the adjacent cluster, until the information reaches the desired destination. In **Figure 6**, nodes 1, 2 and 4 shown by the filled circles are the first hop nodes that successfully decode the message from both the sources, while the nodes that are not able to decode the message from either of the source nodes are shown by hollow circles. The nodes in the first cluster

SNR margin(dB)	0 (%)	10 (%)	15 (%)	20 (%)	25 (%)	30 (%)
D2D Pair 1	572	−6.14	1.58	1.4	0.56	0.19
D2D Pair 2	1258	−3.39	5.15	3.08	1.16	0.39
D2D Pair 3	470	−2.61	5.29	3.10	1.16	0.39
D2D Pair 1, λ = 0.2	39.58	12	5.16	1.86	0.61	0.19
D2D Pair 1, λ = 0.4	93.73	7.27	4.23	1.74	0.60	0.19
D2D Pair 1, λ = 0.6	196.05	2.56	3.33	1.63	0.59	0.19
D2D Pair 1, λ = 0.8	361.02	−1.90	2.45	1.52	0.58	0.19

Table 2. Analysis of RANC D2D with simple relay-aided D2D(including path loss) [27].

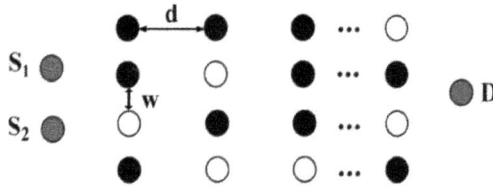

Figure 6. Deterministic network topology: $N = 4$ [28].

employ NC by combining the message from the two sources, i.e., the message I_1 & I_2 transmitted by source S_1 and S_2, respectively, are network coded at the first cluster nodes to form linearly independent codes at each DF relay. Each relay in the first hop transmits these network codes to next cluster of nodes. Each node in the second hop receives network coded copies of I_1 and I_2, which can be decoded using Gaussian elimination. Diversity is achieved at second hop nodes as many network coded copies of I_1 and I_2 are received. The similar process is continued until the transmission are received by the destination. The state of the cluster at each hop is modeled by a Markov chain, as the current state of the system depends only on the previous state of the system.

Random topology In random topology, instead of placing the relay nodes deterministically, the source nodes and the N relay nodes are randomly distributed in a region of area $L \times L$. **Figure 7** illustrates a network where the network is extended in the form of a strip of $L \times L$ sized contiguous regions. A fixed number of nodes at each hop is considered, resulting in a binomial point process (BPP). The transmission flow is similar to deterministic topology, i.e., the nodes that decode the message in the first hop transmit the information to the next hop. The nodes that are able to decode the message from both sources form a codeword. Moreover, in this topology it is assumed that the nodes that decode either I_1 or I_2 are also able to forward the message to the next hop. A single node in a cluster can have four possible states

- State 0 = node does not decode anything,
- State 1 = I_1 is decoded,
- State 2 = I_1 and I_2 are decoded,
- State 3 = I_2 is decoded.

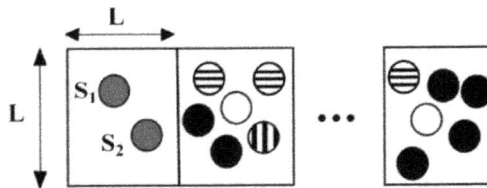

Figure 7. Random topology: Nodes at each hop = 6 [29].

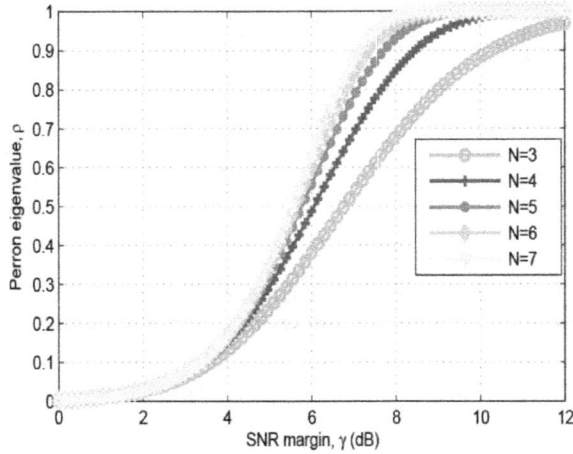

Figure 8. Success probability versus SNR margin for different values of N; $P_t = 1$, $d = 1$ [28].

It is not feasible to model the state of this topology by Markov chain due to the presence of a high number of states. Therefore, state distribution probability model is employed to model the network.

4.2.1. Evaluation model

First, we present results related to the deterministic topology. **Figure 8** shows the relationship between the Perron eigen value, ρ, SNR margin and γ. ρ is the one-hop success probability that corresponds to a state where at least two nodes in a hop are able to successfully decode a message. It can be observed that for a fixed value of N, an increase in the SNR margin leads to an increase in one-hop success probability. Moreover, an increase in the number of nodes, N, leads to higher success probability for a fixed SNR margin because the diversity gain increases. Hence, NC provides a way to transmit data to a far off destination by providing diversity.

The number of hops in the network impact the end-to-end success of data delivery, as the probability of successful hop decreases at each hop. If we want to maintain a certain QoS, η, we need to find the number of nodes in a cluster that would be sufficient for providing the desired QoS. The probability of delivering the message to the mth hop with a constraint that the probability is greater than a threshold η, can be determined through $\rho^m \geq \eta$, where m represents the number of hops. **Figure 9** shows the results pertaining to the normalized distance denoted by $m \times d$ and the QoS. d is distance between two adjacent hops and the results are presented for values of N and η. It can be seen that as the QoS criteria is relaxed, the coverage is extended to a higher normalized distance.

Now, we present the results for random topology. **Figure 10** represents the number of nodes that are in state 2 versus γ, for $N = 8$ and $N = 10$ at the fifth hop. Recall that state 2 is a desired state because a node is in state 2, if it has decoded both I_1 and I_2. The information sent by the

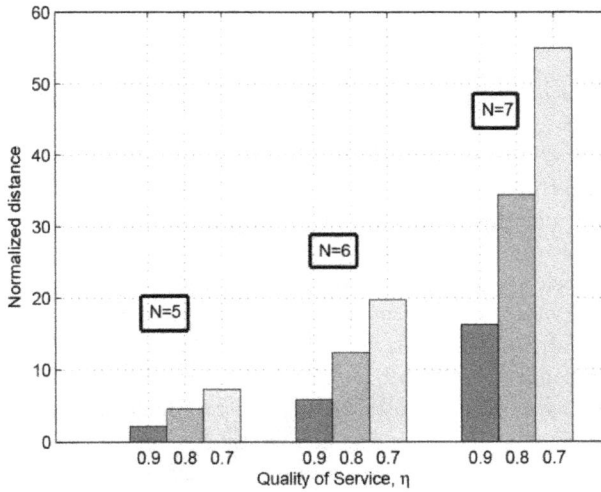

Figure 9. Normalized distance; $\gamma = 6\,dB$, $P_t = 1$, $d = 1$ [28].

sources can be recovered if a minimum of two nodes are available at each hop. The increase in γ leads to an increase in the number of nodes in state 2.

The aforementioned results shown for two multi-hop network topologies empowered by the NC provide a designer's perspective for VMIMO-assisted DAS. The number of nodes that form a part of the VMIMO cluster, and the distance between adjacent clusters impact the network performance. Moreover, NC could help in increasing the end-to-end transmission success probability and therefore increases the coverage area. The network design parameters can be ascertained for VMIMO-assisted DAS to meet the desired QoS criteria.

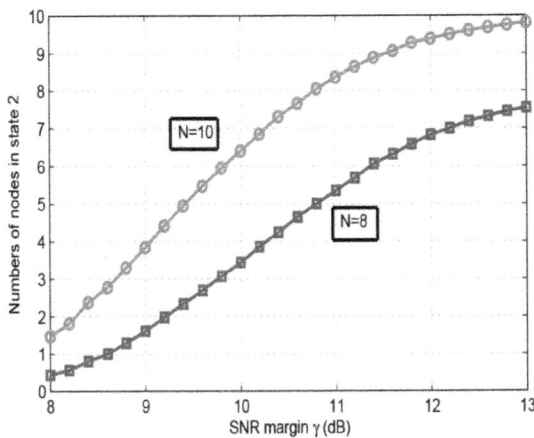

Figure 10. Number of nodes in state 2 versus SNR margin at 5th hop.

5. Conclusion

In this chapter, we presented an overview of the benefits of DAS and the limitations that could arise due to DAS-based network operation. Then, we discussed the concept of VMIMO-assisted DAS and its utility in the network based on single-antenna empowered devices. VMIMO-assisted DAS help in realizing key aspects of 5G technology, i.e., energy/spectral efficiency and enhanced reliability. We analyzed the MIMO-assisted DAS by employing NC and quantify the performance metrics such as end-to-end transmission success probability. We considered a multi-hop environment and based our analysis on two network topologies: (1) Low density DAS (D2D multi-hop networks), and (2) High density DAS (OLA multi-hop networks). We discussed the transmission flow mechanism for both cases and presented results related to network reliability. Moreover, we also quantified the sustainability of the transmissions by determining the maximum distance that could be achieved by operating on particular network parameters. The analysis presented in this chapter provides design insights that could help in identifying the network parameters to achieve the desired QoS. The results highlight the advantages of employing NC in VMIMO-assisted DAS.

Author details

Rafay Iqbal Ansari[1]*, Muhamad Arslan Aslam[2], Syed Ali Hassan[2] and Chrysostomos Chrysostomou[1]

*Address all correspondence to: rafay.ansari@stud.frederick.ac.cy

1 Department of Computer Science and Engineering, Frederick University, Nicosia, Cyprus

2 School of Electrical Engineering and Computer Science (SEECS), National University of Sciences and Technology (NUST), Islamabad, Pakistan

References

[1] Ren H, Liu N, Pan C, Hanzo L. Joint Fronthaul link selection and transmit precoding for energy efficiency maximization of multiuser MIMO-aided distributed antenna systems. IEEE Transactions on Communications. 2017 Dec;65(12):5180-5196

[2] Huang X, Yu R, Kang J, Gao Y, Maharjan S, Gjessing S, et al. Software defined energy harvesting networking for 5G green communications. IEEE Wireless Communications. 2017 Aug;24(4):38-45

[3] Ren H, Liu N, Pan C, He C. Energy efficiency optimization for MIMO distributed antenna systems. IEEE Transactions on Vehicular Technology. 2017 March;66(3):2276-2288

[4] Wang J, Wang Y, Feng W, Su X, Zhou S. An iterative power allocation scheme for improving energy efficiency in massively dense distributed antenna systems. In: 2016 IEEE 83rd Vehicular Technology Conference (VTC Spring); 2016. pp. 1-5

[5] Bengtsson EL, Rusek F, Malkowsky S, Tufvesson F, Karlsson PC, Edfors O. A simulation framework for multiple-antenna terminals in 5G massive MIMO systems. IEEE Access. 2017;**5**:26819-26831

[6] Soorki MN, Manshaei MH, Maham B, Saidi H. On uplink virtual MIMO with device relaying cooperation enforcement in 5G networks. IEEE Transactions on Mobile Computing. 2018 Jan;**17**(1):155-168

[7] Yang T, Sun QT, Zhang JA, Yuan J. A linear network coding approach for uplink distributed MIMO systems: Protocol and outage behavior. IEEE Journal on Selected Areas in Communications. 2015 Feb;**33**(2):250-263

[8] Chen X, Guo S, Wu Q. Link-level analysis of a multiservice indoor distributed antenna system [wireless corner]. IEEE Antennas and Propagation Magazine. 2017 June;**59**(3):154-162

[9] Wang HH, Hou HA, Chen JC. Design and analysis of an antenna control mechanism for time division Duplexing distributed antenna systems over high-speed rail communications. IEEE Transactions on Emerging Topics in Computing. 2016;**4**(4):516-527

[10] Feng W, Chen Y, Shi R, Ge N, Lu J. Exploiting macrodiversity in massively distributed antenna systems: A controllable coordination perspective. IEEE Transactions on Vehicular Technology. 2016 Oct;**65**(10):8720-8724

[11] Ge X, Jin H, Leung VCM. Opportunistic downlink scheduling with resource-based fairness and feedback reduction in distributed antenna systems. IEEE Transactions on Vehicular Technology. 2016 July;**65**(7):5007-5021

[12] He C, Li GY, Zheng FC, You X. Power allocation criteria for distributed antenna systems. IEEE Transactions on Vehicular Technology. 2015 Nov;**64**(11):5083-5090

[13] Heath RW, Wu T, Kwon YH, Soong ACK. Multiuser MIMO in distributed antenna systems. In: 2010 Conference Record of the Forty Fourth Asilomar Conference on Signals, Systems and Computers; 2010. pp. 1202-1206

[14] Lu L, Li GY, Swindlehurst AL, Ashikhmin A, Zhang R. An overview of massive MIMO: Benefits and challenges. IEEE Journal of Selected Topics in Signal Processing. 2014 Oct;**8**(5):742-758

[15] Li Q, Li G, Lee W, i Lee M, Mazzarese D, Clerckx B, et al. MIMO techniques in WiMAX and LTE: A feature overview. IEEE Communications Magazine. 2010 May;**48**(5):86-92

[16] Castaeda E, Silva A, Gameiro A, Kountouris M. An overview on resource allocation techniques for multi-user MIMO systems. IEEE Communications Surveys Tutorials. 2017 Firstquarter;**19**(1):239-284

[17] Ji H, Kim Y, Lee J, Onggosanusi E, Nam Y, Zhang J, et al. Overview of full-dimension MIMO in LTE-advanced pro. IEEE Communications Magazine. 2017 February;**55**(2):176-184

[18] Fodor G, Rajatheva N, Zirwas W, Thiele L, Kurras M, Guo K, et al. An overview of massive MIMO technology components in METIS. IEEE Communications Magazine. 2017;**55**(6):155-161

[19] Karakus C, Diggavi S. Enhancing multiuser MIMO through opportunistic D2D coopera-tion. IEEE Transactions on Wireless Communications. 2017 Sept;**16**(9):5616-5629

[20] Kwon B, Lee S. Effective interference Nulling virtual MIMO broadcasting transceiver for multiple relaying. IEEE Access. 2017;**5**:20695-20706

[21] Zhang M, Wen M, Cheng X, Yang L. A dual-hop virtual MIMO architecture based on hybrid differential spatial modulation. IEEE Transactions on Wireless Communications. 2016 Sept;**15**(9):6356-6370

[22] Rodziewicz M. Network coding aided device-to-device communication. In: Wireless Conference (European Wireless), 2012 18th European. VDE; 2012. pp. 1-5

[23] Jayasinghe LS, Jayasinghe P, Rajatheva N, Latva-aho M. MIMO physical layer network coding based underlay device-to-device communication. In: Personal Indoor and Mobile Radio Communications (PIMRC), 2013 IEEE 24th International Symposium on. IEEE; 2013. pp. 89-94

[24] Pahlevani P, Hundebøll M, Pedersen M, Lucani D, Charaf H, Fitzek FP, et al. Novel con-cepts for device-to-device communication using network coding. IEEE Communications Magazine. 2014;**52**(4):32-39

[25] Pyattaev A, Galinina O, Andreev S, Katz M, Koucheryavy Y. Understanding practical limitations of network coding for assisted proximate communication. IEEE Journal on Selected Areas in Communications. 2015;**33**(2):156-170

[26] Wei L,Wu G, Hu RQ. Multi-pair device-to-device communications with spacetime analog network coding. In: Wireless Communications and Networking Conference (WCNC), 2015 IEEE. IEEE; 2015. pp. 920-925

[27] Ansari RI, Hassan SA, Chrysostomou C. RANC: Relay-aided network-coded D2D network. In: Information, Communications and Signal Processing (ICICS), 2015 10th International Conference on. IEEE; 2015. pp. 1-5

[28] Aslam MA, Hassan SA. Analysis of multi-source multi-hop cooperative networks employing network coding. In: Vehicular Technology Conference (VTC Spring), 2015 IEEE 81st. IEEE; 2015. pp. 1-5

[29] Aslam MA, Hassan SA. Coverage analysis of a dual source opportunistic network utiliz-ing cooperation. In: Information, Communications and Signal Processing (ICICS), 2015 10th International Conference on. IEEE; 2015. pp. 1-5

Chapter 4

Network Coding-Based Next-Generation IoT for Industry 4.0

Goiuri Peralta, Raul G. Cid-Fuentes, Josu Bilbao and
Pedro M. Crespo

Additional information is available at the end of the chapter

http://dx.doi.org/10.5772/intechopen.78338

Abstract

Industry 4.0 has become the main source of applications of the Internet of Things (IoT), which is generating new business opportunities. The use of cloud computing and artificial intelligence is also showing remarkable improvements in industrial operation, saving millions of dollars to manufacturers. The need for time-critical decision-making is evidencing a trade-off between latency and computation, urging Industrial IoT (IIoT) deployments to integrate fog nodes to perform early analytics. In this chapter, we review next-generation IIoT architectures, which aim to meet the requirements of industrial applications, such as low-latency and highly reliable communications. These architectures can be divided into IoT node, fog, and multicloud layers. We describe these three layers and compare their characteristics, providing also different use-cases of IIoT architectures. We introduce network coding (NC) as a solution to meet some of the requirements of next-generation communications. We review a variety of its approaches as well as different scenarios that improve their performance and reliability thanks to this technique. Then, we describe the communication process across the different levels of the architecture based on NC-based state-of-the-art works. Finally, we summarize the benefits and open challenges of combining IIoT architectures together with NC techniques.

Keywords: industry 4.0, IoT, IIoT, latency, network coding, fog computing, multicloud

1. Introduction

Industry 4.0, that is, the fourth industrial revolution, represents industry and manufacturing digitalization bringing with it, among other things, the so-called smart factories. This transformation comes through the adoption of the Internet of Things (IoT) [1], which gives rise to the Industrial IoT (IIoT) and allows to interconnect humans, machines, and smart devices, as well as to share huge amounts of data among them.

In order to cope with big data and predictive analytics [2], cloud computing is becoming another key enabler due to its computing, storage, and networking capabilities. It allows us to obtain meaningful information and valuable insights which will increase the efficiency, productivity, and performance of manufacturing processes and services. Several IIoT applications, such as system control, anomaly detection, or robot guidance, are time-critical, and therefore, they require millisecond response times. Thus, low-latency communications, as well as real-time analysis and monitoring, are indispensable for immediate decision-making.

Although the cloud offers high scalability, flexibility, and responsiveness, cloud-based analytics may introduce excessive latency, which would compromise the performance of time-critical applications. In order to accomplish a trade-off between latency and computation, IIoT deployments are moving cloud capabilities downwards to fog nodes to perform early analytics and minimize latency. Furthermore, most delay-critical applications not only require low-latency communications but also ensure high reliability. A promising technique that increases network reliability while reducing end-to-end latency is network coding (NC). Its properties are particularly beneficial for enhancing the robustness and reducing delays of wireless sensor network (WSN) communications [3]. Moreover, it improves the efficiency of distributed storage systems, regarding both data download speed and redundancy [4].

In this chapter, we overview next-generation IIoT systems, which must provide low-latency communications as well as ensure their reliability in order to allow the performance of on-premise advanced cloud analytics for time-critical IIoT applications, that is, to bring the cloud to the fog (see **Figure 1**). This objective can be achieved by implementing a three-layer architecture based on IoT nodes, fog nodes and a multicloud environment, and also by exploiting the advantageous properties of NC techniques across the architecture.

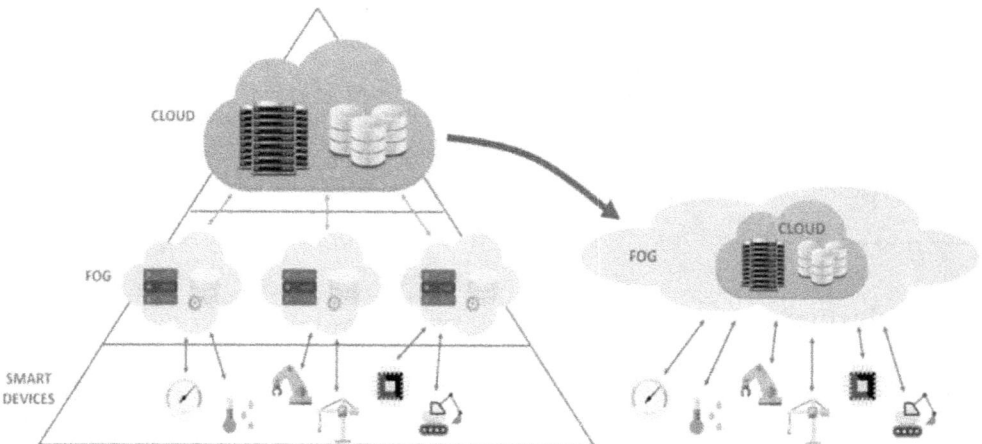

Figure 1. Bringing the cloud to the fog.

The rest of this chapter is organized as follows. First, we overview next-generation IIoT architectures, briefly describing and comparing the different layers as well as providing different use-cases in which these architectures are integrated. Next, we introduce some NC approaches and describe the benefits of NC regarding different scenarios. Then, we describe the communication process across the different levels of the architecture. We also summarize the benefits of merging IIoT architectures and NC techniques. Finally, we discuss existing issues and open challenges, and we report the final conclusions of the chapter.

2. Next-generation IIoT architectures

Nowadays, due to its scalability and big data management capabilities, cloud-based architectures are most widely used in Industry 4.0 applications. However, the integration of the IoT into industrial environments poses new challenges, which implies an architectural adaptation. As previously mentioned, IIoT applications are mostly delay-sensitive and require instant decision-making. This has led to the integration of fog nodes into the industrial systems in order to perform early analytics and closed-loop control. Moreover, systems of this nature must be robust. Thus, with the aim of providing a fault-tolerant architecture and guarantee system reliability, multicloud deployments are emerging as a promising solution. In addition to the latter, they enable to use the connections under the best conditions and therefore, delays can be reduced. Dependencies on a single cloud provider can also be avoided.

It can be said that next-generation IIoT architectures, as shown in **Figure 2**, will consist of three layers, composed of IoT or smart devices, fog nodes and multiple clouds. The lowest layer, comprised of a variety of end-nodes, is responsible for sending taken measurements to actuators or fog devices. In the fog layer, time-critical analytics, as well as closed-loop control, can be performed. Finally, cloud servers are in charge of heavy data analytics and compute-intense workloads that manage a vast amount of data.

2.1. Architecture design

A description and comparison of the layers that comprise next-generation IIoT architectures are next provided.

2.1.1. WSN

WSNs can be considered the main communication technologies of IIoT due to the flexibility they offer to connect and manage a large number of sensors and actuators, independently of their location. A WSN consists of several IoT nodes, including sensors, actuators, and smart devices, which take several measurements. These devices are mainly battery, storage, and processing power constrained. This layer is responsible for gathering sensor data, such as machine temperature or vibration measurements, and for uploading them. It also receives instructions from the upper layers in order to perform a corresponding task or action.

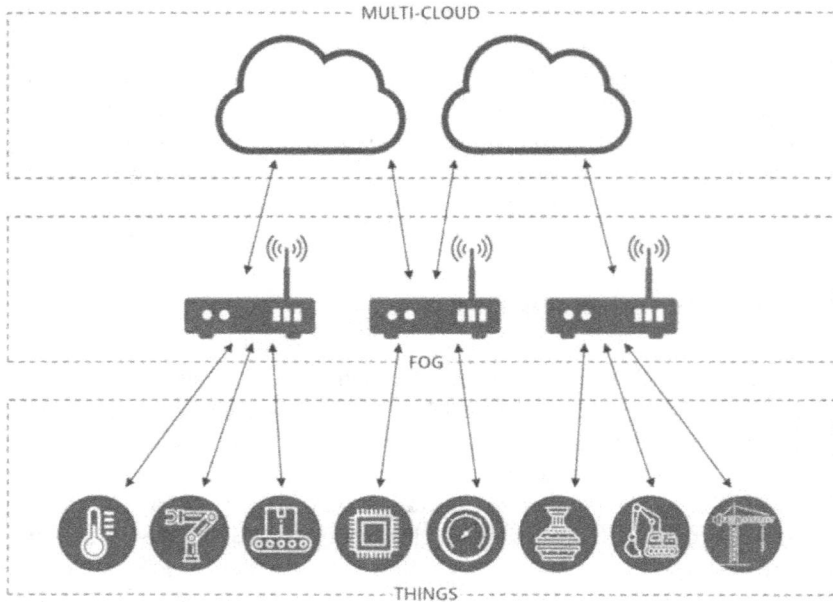

Figure 2. Next-generation IIoT architecture.

2.1.2. Fog

The fog can be considered as an intermediate layer between the cloud and IoT devices and so, it extends cloud computing capabilities to the edge of the network [5]. One of its main advantages is its closeness to the end-nodes, which makes possible to reduce communication latency and to enable real-time service support. Since fog computing allows early data processing, the amount of data sent to the cloud can be reduced. In addition, its mobility and location-awareness enable to deliver rich services to moving devices [6].

2.1.3. Multicloud

The cloud can be described as several distributed remote servers which can be accessed via the Internet to store and manage big amounts of data [7]. Cloud computing enables the remote on-demand use of computing resources, that is, networks, servers, storage, applications, and services. It provides virtualized, elastic, and controllable services and powerful computational capabilities, enabling complex application systems at lower costs. The deployment of more than one cloud, in addition to the mentioned advantages, provides fault tolerance against service outages, and the system security level is improved since it is possible to store the information divided into different clouds. Furthermore, application requirements can be better adapted to available cloud resources and connectivity conditions.

2.1.4. Comparison

With the adoption of IoT, the number of things connected to the Internet is expected to grow up to 20 billion in 2020 [8]. Thus, a scalable architecture is required in order to adapt to such a huge number of devices. Moreover, the need for large amounts of data to be accessed more quickly is ever-increasing, where the inherent latency of the cloud can be detrimental. Latency issues become highly damaging, particularly for IIoT time-critical applications. Autonomous decisions are required in order to prevent failures or optimize production, and thus, milliseconds matter when trying, for instance, to prevent manufacturing line downtimes or to get the right decision in autonomous vehicles.

Processing data directly in the end-devices would be the best solution in order to provide the lowest latency and jitter. However, the constrained nature of these nodes inhibits the performance of more advanced processing and analytics. Thus, fog computing can be the most suitable solution for applications that cannot afford the delay caused by the round trip to the cloud server. Nonetheless, fog computing requires local management of redundancy and data backup. Moreover, the integration of devices capable of performing remote data analytics implies the increase of the architecture complexity, as well as of the associated costs in hardware and software investments. **Table 1** shows the most significant differences between WSNs, fog computing, and cloud computing.

2.2. Application use-cases

The three-layer architecture enables to exploit the efficiency and scalability of the fog while benefiting from the powerful storage and computing resources of the cloud. Next, we show some use-case examples.

2.2.1. Smart energy

Wind energy-based smart grids require data analysis and real-time decision making. In a large wind farm, the health of the turbines is monitored by analyzing data collected by

Feature	WSNs	Fog computing	Cloud computing
Latency	Very low	Low	High
Delay jitter	Very low	Low	High
Server location	—	Local	Internet
Client–server distance	—	One hop	Multiple hops
Location awareness	Yes	Yes	No
Distribution	Highly distributed	Distributed	Centralized
Mobility awareness	Guaranteed	Supported	Limited
Real-time interactions	Guaranteed	Supported	Limited

Table 1. Comparison between WSN, fog, and cloud computing [9].

numerous sensors [10]. Each turbine can be monitored locally, that is, in the fog, and the collective performance can be improved by processing data on remote servers in the cloud. Thus, it enables to combine real-time response for early actions and advanced analyses for a deeper view of the whole wind farm. It can increase energy output, decrease operational costs, and increase turbine uptime.

2.2.2. Smart transportation

Self-driving vehicles, for instance, are equipped with an on-board system that, through real-time data analysis, allows controlling the car without human interaction. In such systems, highly reliable and low latency communication is crucial. Thus, critical decisions that require an instantaneous response are better managed with fog computing [11]. However, for monitoring the tracking performance of a truck fleet [12], as there is no need for real-time analytics, cloud computing is more suitable. Advanced cloud analytics based on information gathered from different parts of the truck can bring insights to improve the maintenance and lower repair costs.

2.2.3. Smart manufacturing

Smart factories are able to perform predictive maintenance of their machines or improve product quality by real-time sensor analysis [13]. Fog computing is crucial for these delay-critical data processing. However, cloud computing can provide an overall system management as well as machine learning analytics that require greater computing power.

2.2.4. Smart cities

Fog computing can provide a fast, real-time, and location-aware solution for many IoT use cases of smart cities, such as smart buildings [14]. Several sensors gather diverse measurements like temperature, energy usage, humidity, parking occupancy, air quality, elevators, smoke, and so on. The efficiency of the system can be improved by managing critical data at the fog layer in real time, as in traffic control, and by performing big data analytics in the cloud.

3. Next-generation communications

Most IIoT systems are deployed in harsh environments, where different devices within the architecture can be connected and disconnected from the network any time. Thus, besides providing low-latency communications, it is crucial to strengthen these communications in order to ensure a robust and highly reliable environment.

3.1. Issues

IIoT networks require system reliability, data availability and high communication quality. This may be difficult to achieve due to inherent constraints of these scenarios. WSNs, for example, may suffer from noise or multipath interferences, among others, which cause packet loss and inevitably degrades the quality of the communications. Moreover, the dynamic topology of these architectures in which devices connect intermittently, can destabilize communications and introduce variable delays. In order to overcome these issues, the integration of NC techniques across the shown architecture can be a suitable solution due to its properties.

3.2. Network coding

NC breaks with the traditional store-and-forward transmission model [15] by allowing any intermediate node to recombine incoming packets into coded ones, which are decoded at destination. Its properties make it a promising solution to improve the performance of wireless and peer-to-peer networks. It exploits the broadcast nature of the wireless medium [3], which facilitates node cooperation to provide significant benefits in terms of communication robustness, stability, throughput, and latency.

Moreover, dependency on obtaining a particular packet is removed by applying NC since it is sufficient to get enough linearly independent packet combinations in order to recover the required data. Thus, in distributed systems, such as P2P [16] or multicloud environments [17], the use of NC can reduce additional data download or access delays in highly loaded conditions as well as improve the performance of data recovery and acquisition [4].

3.2.1. Advanced techniques

Advanced NC techniques are based on the widely used NC approach random linear network coding (RLNC) [18], where the received K packets are linearly combined with randomly chosen coefficients from a finite field or Galois Field (GF) F_q before forwarding them. The performance of RLNC is influenced by the following coding parameters: finite field size, generation size, and coding vector density [19], among others. Eq. (1) represents the encoding process, where coding coefficients ($a_{i,j} \in F_q$) multiply incoming packets $M_i (\forall i = 1, ..., K)$ to generate corresponding encoded packets X_i.

$$\begin{bmatrix} X_1 \\ X_2 \\ \vdots \\ X_K \end{bmatrix} = \begin{bmatrix} \alpha_{1,1} & \alpha_{1,2} & \cdots & \alpha_{1,N} \\ \alpha_{2,1} & \alpha_{2,2} & \cdots & \alpha_{2,N} \\ \vdots & \vdots & \ddots & \vdots \\ \alpha_{K,1} & \alpha_{K,2} & \cdots & \alpha_{K,N} \end{bmatrix} \cdot \begin{bmatrix} M_1 \\ M_2 \\ \vdots \\ M_K \end{bmatrix} \tag{1}$$

Several variants of this technique have been developed in order to adapt it to different scenarios and application requirements. Perpetual codes [20], for instance, can be considered as a supplement of RLNC. In manifold scenarios, particularly for large generation sizes, they can substantially increase the throughput due to their sparsity and the possibility of structured decoding.

For heterogeneous networks with devices of different resources, fulcrum codes [21] allow to use binary GF operations in the network to achieve reduced overhead and computational cost, and reach compatibility with heterogeneous devices and data flows in the network, while providing the opportunity of employing higher coding finite fields end-to-end for greater performance. On the other hand, systematic coding [22] allows sending coded packets along with original ones, that is, uncoded packets, which can help to reduce overhead and improve the real-time decoding performance.

For low-delay applications such as real-time control applications, on-the-fly or sliding window coding can be the most suitable solutions. Unlike block codes where all packets in a block need to be present to start generating useful coded packets, on-the-fly codes [23] are able to encode data while they become available and these packets are progressively decoded. Sliding window codes [24] are more flexible since they remove the limitation of fixed blocks by creating a variable-sized sliding window.

Finally, Tunable Sparse Network Coding (TSNC) [25], unlike RLNC that applies a fixed coding density for the entire process, tunes the density of coded packets during transmission to adjust to the trade-off between real-time performance and reliability. TSNC proposes to increase the coding density as the destination node receives more linearly independent packets since the probability of receiving innovative packets is lower, thereby reducing coding complexity.

3.2.2. NC benefits

This section lists different benefits of NC, showing its suitability for IIoT systems and applications.

- *Distributed storage systems*: IIoT systems require high reliability and availability. Thus, data must be distributed and stored in such a way as to ensure fault tolerance, for example in the event of a server failure. Packet loss, delay, and bandwidth fluctuation can hinder data distribution. The main benefit of NC over P2P environments is in relation to the coupon collector problem [16], being able to solve this issue due to the redundancy introduced in packet transmissions. Therefore, the performance of data streaming is enhanced since download times are minimized. With NC, the performance of the system depends much less on the underlying topology and schedule.

  NC can help to increase the reliability of distributed storage systems like multicloud deployments [17]. In case of data loss, the amount of redundant data required for repair is minimized. In addition, each cloud is used at its maximum speed even in highly loaded conditions or dynamically changing environments. Thus, NC improves storage efficiency in terms of data retrieval time and storage space.

- *Dynamic topologies*: NC techniques can be helpful for efficient content distribution [26] in changing environments. For example in Vehicular Ad-Hoc Networks (VANETs), in order to avoid possible accidents, vehicles exchange road state information among them. Even in dynamic road changing conditions, VANET applications, such as traffic live video broadcast, must guarantee a correct data reception. Since NC enhances network performance and reduces the number of required data transmissions, it can reduce transmission delays.

  Due to the previous and together with its decentralized nature and robustness, NC can be extrapolated also to dynamic IIoT architectures, where end-nodes may connect periodically in order to save power or they can connect to different access points.

- *Constrained environments*: the use of NC has been extended also to constrained environments. In satellite communications, for instance, bandwidth is usually limited and round-trip delays are high. The properties of this technique can be advantageous particularly in multibeam satellites [27]. On the one hand, NC improves throughput and bandwidth usage. On the other hand, it does not require any change at the physical layer and thus, it easies the implementation on already deployed satellite systems. NC techniques can also be used in applications aimed at energy-efficient data transmissions, such as Wireless Body Area Networks (WBANs), since they can provide reliable communications under low-energy constraints [28]. Therefore, IIoT applications can also profit from this technique as the majority of end-devices are resource constrained.

- Poor quality channels: NC can improve the transmission performance in environments with unstable channel conditions which quality may not meet end-user quality of service (QoS) requirements, such as delay and reliability. An example of the previous are Underwater Sensor Networks (UWSNs) and Power Line Communication (PLC) systems. In UWSNs, the acoustic communications suffer high error rates and long propagation delays, which require efficient error recovery. NC can exploit the broadcast property of acoustic channels, improving data throughput [29]. PLC systems, on the other hand, are able to provide multicast and broadcast services by exploiting existing electrical wires. Due to the similarities between power line and wireless channels, NC protocols can be applied in order to achieve the implementation constraints [30] and provide reliable communications in harsh environments.

IIoT systems that relay on WSNs may deal with interferences or channel contention that cause QoS issues. Thus, NC-based techniques can help to improve channel resources as well as data rate while maintaining QoS.

4. NC over IIoT architectures

In this section, we introduce NC into next-generation IIoT architectures reviewing related state-of-the-art works. We also outline some of the most relevant benefits and challenges.

4.1. Communication process

In IIoT systems, not only low-latency communications between end-nodes (things) and the cloud must be guaranteed but the whole system must be robust, including the connections and the provided service. Next, the communication process across the architecture is described.

4.1.1. Things

Implementing NC techniques through the WSN, communication latency can be reduced [31] and its robustness [32, 33] improved. Here, sensors and actuators combine their measurements and transmit them across the network. By using NC, intermediate nodes recode received data and send them to one or more gateways which compose the fog layer. These devices are then in charge of uploading incoming data to the multicloud framework.

4.1.2. Fog

Fog nodes can be any device with computing, storage, and network connectivity, such as controllers, routers, gateways, and so on. They can be deployed anywhere with a network connection, for instance, alongside a factory floor. They are interconnected among them, with the IoT devices and also with cloud servers, forming a distributed network. Therefore, NC-based techniques can be extrapolated from WSNs to the communication between the devices within the fog layer [34, 35].

The fog layer is responsible for gathering data from end-devices and for distributing coded packets to the different clouds that comprise the multicloud deployment. The use of NC has also been demonstrated to be beneficial for data distribution [36]. Moreover, this technique is advantageous for distributed storage systems, since it can achieve an optimal trade-off between storage and repair traffic. Thus, it can also help to deal with fog storage nodes that may continuously leave the network without a replacement [37].

4.1.3. Multicloud

Clouds within the multicloud deployment are responsible for storing incoming network-coded data from the lower layer. NC-based techniques can improve the process of lost data recovery, as well as enhance the efficiency of data redundancy [38]. As an example, **Figure 3** illustrates the repair operation in case of a cloud failure using exact minimum-storage regenerating (EMSR) codes. A file is divided into for fragments, and both original and coded chunks are distributed as shown in the figure. Assuming Cloud 1 is down (A and B are lost), the surviving nodes XOR their own chunks to create new encoded ones in order to make possible the reconstruction of A and B.

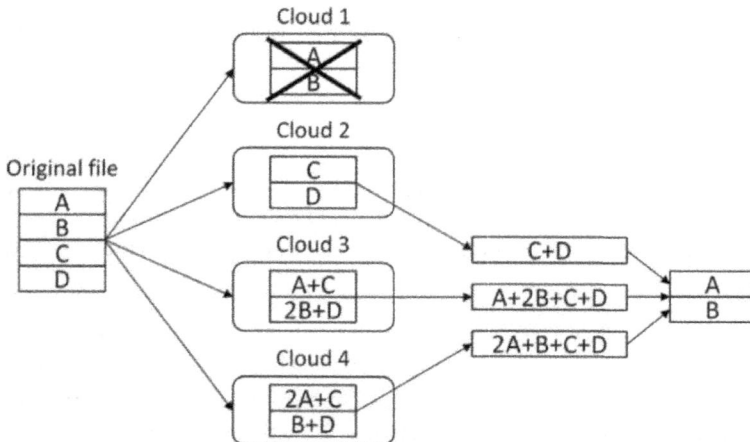

Figure 3. Repair process with EMSR codes [17].

Clouds are then able to perform required operations for decision-making or further analytics. Data or action commands are transmitted to the corresponding devices based on the results of the performed analysis. The cloud response time can also be reduced, as in the upstream communication, due to the implementation of NC. In [39], for instance, if a file has been divided into g fragments, the number of coded packets P_i stored on each of the N clouds is calculated in order to achieve optimal scheduling during data retrieval (see Eqs. (2) and (3), where R_i defines the download rate and α_i denotes the minimum part of the data to be stored).

$$\alpha_i = \frac{R_i}{\sum_{i=1}^{N} R_i} \tag{2}$$

$$P_i \geq \alpha_i \cdot g \tag{3}$$

If the information is stored over distributed untrusted platforms, such as public clouds, the inherent use of NC schemes can provide, in addition to fault tolerance, a security level against eavesdroppers [40]. However, it is necessary to find a trade-off between fault-tolerance and security [41], since the more redundant data, the more vulnerable the system becomes.

4.2. Summary

We overview the most relevant benefits the integration of IIoT architectures and NC provides. While this approach can bring significant advantages, it also poses some issues. We identify and describe future challenges that may arise with the implementation of next-generation IIoT architectures.

4.2.1. Benefits

As the IIoT architecture relies on a multicloud deployment, the reliability and availability of the entire system can be enhanced. Data are distributed across different clouds, and so, the possibilities of suffering a cyber-attack are reduced. Moreover, this information is stored differently from the original form. Thus, data privacy is improved. The multicloud environment, due to its fault tolerance, increases the robustness in case of service outages. Due to the implementation of multiple clouds, this architecture enables to distribute data to the most convenient cloud, which makes possible not only to choose the service provider that better fits the moment requirements but to use the connections under the best conditions. Thus, it helps to identify the right service architecture to optimize latency, location, and cost.

The use of NC-based techniques can enhance the performance of the communications over congested WSNs, as well as of the data distribution and recovery processes over multicloud deployments. Since data redundancy is more efficient, reliability and availability of the provided service are improved. Besides, the integration of NC-based techniques into the architecture can lead to the reduction of end-to-end latency.

All of the above advantages allow the computational power of the cloud to be available closer to the end nodes, improving the performance of delay-sensitive IIoT applications.

4.2.2. Challenges

As mentioned throughout the chapter, next-generation IIoT architectures aim to reduce the end-to-end communication latency and to increase the system reliability by merging both fog and multicloud-based schemes as well as NC techniques. However, the use of complex NC schemes can result in extra delays taking into account that IoT devices have limited computational resources. Thus, in order to exploit the benefits of this technique, it is crucial to choose the most suitable coding parameters as well as to design simpler coding schemes and adaptable scheduling and routing algorithms.

IIoT systems must also provide scalability and flexibility. A cloud environment is inherently a scalable architecture due to its capability to manage network topology variations while handling big amounts of data. However, in architectures such as the proposed, as devices may be intermittently connected to the network, not only the architecture itself needs to be scalable and adaptable to changing environments, but also the coding techniques.

5. Conclusions

This chapter overviews next-generation IIoT systems which, in order to satisfy the demands of Industry 4.0 applications, must ensure low-latency and highly reliable communications. This will enable advanced analytics for time-critical IIoT applications. The previous objective can be achieved on one hand, by implementing a three-layer architecture based on IoT devices, fog nodes, and a multicloud deployment. On the other hand, the use of NC techniques across this architecture can improve the communication quality and increase the system reliability. In this chapter, we describe next-generation IIoT architectures and provide different application use-cases where they can be applied. We also review NC-based techniques and the benefits of this technique for different scenarios. Next, we describe the introduction of NC for the communications across the architecture. We also outline the advantages of the approach and finally, we present some challenges that may arise, such as the design of scalable and adaptive coding schemes and routing algorithms, and which may inspire future research lines.

Acknowledgements

This work has been partially supported by the Basque Government through the Elkartek program, as well as the Spanish Ministry of Economy and Competitiveness through the CARMEN project (TEC2016-75067-C4-3-R) and the COMONSENS network (TEC2015-69648-REDC) and the H2020 research framework of the European Commission.

Author details

Goiuri Peralta[1]*, Raul G. Cid-Fuentes[1], Josu Bilbao[1] and Pedro M. Crespo[2]

*Address all correspondence to: gperalta@ikerlan.es

1 IK4-Ikerlan Technology Research Center, Arrasate-Mondragón, Spain

2 University of Navarra (Tecnun), Donostia-San Sebastián, Spain

References

[1] Da Xu L, He W, Li S. Internet of things in industries: A survey. IEEE Transactions on Industrial Informatics. 2014;**10**(4):2233-2243. DOI: 10.1109/TII.2014.2300753

[2] Yue X, Cai H, Yan H, Zou C, Zhou K. Cloud-assisted industrial cyber-physical systems: An insight. Microprocessors and Microsystems. 2015;**39**(8):1262-1270. DOI: 10.1016/j.micpro.2015.08.013

[3] Fragouli C, Boudec JYL, Widmer J. Network coding: An instant primer. ACM SIGCOMM Computer Communication Review. 2006;**36**(1):63-68. DOI: 10.1145/1111322.1111337

[4] Dimakis AG, Godfrey PB, Wu Y, Wainwright MJ, Ramchandran K. Network coding for distributed storage systems. IEEE Transactions on Information Theory. 2010;**56**(9):4539-4551. DOI: 10.1109/TIT.2010.2054295

[5] Bonomi F, Milito R, Zhu J, Addepalli S. Fog computing and its role in the internet of things. In: Proceedings of the First Edition of the MCC Workshop on Mobile Cloud Computing (MCC '12); 17 August 2012. ACM; 2012. pp. 13-16

[6] Peralta G, Iglesias-Urkia M, Barcelo M, Gomez R, Moran R, Bilbao J. Fog computing based efficient IoT scheme for the industry 4.0. In: International Workshop of Electronics, Control, Measurement, Signals and their Application to Mechatronics (ECMSM); 24-26 May 2017. IEEE; 2017. pp. 1-6

[7] Zhang Q, Cheng L, Boutaba R. Cloud computing: State-of-the-art and research challenges. Journal of Internet Services and Applications. 2010;**1**(1):7-18. DOI: 10.1007/s13174-010-0007-6

[8] Gartner Inc. Analysts to Explore the Value and Impact of IoT on Business. Gartner Symposium/ITxpo [Internet]. 2015. Available from: https://www.gartner.com/news-room/id/3165317 [Accessed: January 10, 2018]

[9] Cisco Blog. IoT, from Cloud to Fog Computing [Internet]. 2015. Available from: https://blogs.cisco.com/perspectives/iot-from-cloud-to-fog-computing [Accessed: January 05, 2018]

[10] FOGHORN. Wind Turbine Optimization [Internet]. 2018. Available from: https://www.foghorn.io/wind-turbine-optimization/ [Accessed: January 10, 2018]

[11]   Automation Alley. Fog computing: A New Paradigm for the Industrial IoT [Internet]. Available from: https://www.automationalley.com/Blog/October-2017/Fog-Computing-A-New-Paradigm-for-the-Industrial-Io.aspx [Accessed: January 05, 2018]

[12]   Thorn Technologies. How Edge Computing and the Cloud will Power the Future of IoT [Internet]. Available from: https://www.thorntech.com/2017/11/edge-computing-and-the-cloud-future-of-iot/ [Accessed: January 10, 2018]

[13]   Fog in the Factory [Internet]. Available from: https://industrial-iot.com/2017/01/fog-in-the-factory/ [Accessed: January 05, 2018]

[14]   Dutta J, Sarbani R. IoT-fog-cloud based architecture for smart city: Prototype of a smart building. In: 7th International Conference on Cloud Computing, Data Science & Engineering-Confluence; 12-13 January, 2017. IEEE; 2017. pp. 237-242

[15]   Ahlswede et al. Network information flow. IEEE Transactions on Information Theory. 2000;**46**(4):1204-1216. DOI: 10.1109/18.850663

[16]   Wang M, Li B. Network coding in live peer-to-peer streaming. IEEE Transactions on Multimedia. 2007;**9**(8):1554-1567. DOI: 10.1109/TMM.2007.907460

[17]   Chen HCH, Hu Y, Lee PPC, Tang Y. NCCloud: A network-coding-based storage system in a cloud-of-clouds. IEEE Transactions on Computers. 2014;**63**(1):31-44. DOI: 10.1109/TC.2013.167

[18]   Ho et al. A random linear network coding approach to multicast. IEEE Transactions on Information Theory. 2006;**52**(10):4413-4430. DOI: 10.1109/TIT.2006.881746

[19]   Heide J, Pedersen MV, Fitzek FHP, Medard M. On code parameters and coding vector representation for practical RLNC. In: IEEE International Conference on Communications (ICC); 5-9 June 2011. IEEE; 2011. pp. 1-5

[20]   Heide J, Pedersen MV, Fitzek FH, Médard M. A perpetual code for network coding. In: 79th Vehicular Technology Conference (VTC Spring); 18-21 May 2014. IEEE; 2015. pp. 1-6

[21]   Lucani DE, Pedersen MV, Ruano D, Sørensen CW, Fitzek FH, Heide J, et al. Fulcrum network codes: A code for fluid allocation of complexity. 2014; arXiv:14046620

[22]   Shrader B, and Jones NM. Systematic wireless network coding. In: Military Communications Conference (MILCOM); 18-21 October 2009. IEEE; 2010. pp. 1-7

[23]   Tournoux PU, Lochin E, Lacan J, Bouabdallah A, Roca V. On-the-fly erasure coding for real-time video applications. IEEE Transactions on Multimedia. 2011;**13**(4):797-812. DOI: 10.1109/TMM.2011.2126564

[24]   Li B, Bi S, Zhang R, Jiang Y, Li Q. Random network coding based on adaptive sliding window in wireless multicast networks. In: 83rd Vehicular Technology Conference (VTC Spring); 15-18 May 2016. IEEE; 2016. pp. 1-5

[25]   Feizi S, Lucani DE, Sørensen CW, Makhdoumi A, Médard M. Tunable sparse network coding for multicast networks. In: International Symposium on Network Coding (NetCod); 27-28 June 2014. IEEE; 2014. pp. 1-6

[26] Ahmed S, Kanhere SS. VANETCODE: Network coding to enhance cooperative downloading in vehicular ad-hoc networks. In: Proceedings of the 2006 International Conference on Wireless Communications and Mobile Computing (IWCMC); 3 July 2006. ACM; 2006. pp. 527-532

[27] Vieira F, Lucani DE, Alagha N. Codes and balances: Multibeam satellite load balancing with coded packets. In: IEEE International Conference on Communications (ICC); 10-15 June 2012. Ottawa: IEEE; 2012. pp. 3316-3321

[28] Arrobo GE, Gitlin RD. Improving the reliability of wireless body area networks. In: Annual International Conference of the IEEE Engineering in Medicine and Biology Society (EMBC'11); 30 August–3 September 2011. Boston: IEEE; 2011. pp. 2192-2195

[29] Wu H, Chen M, Guan X. A network coding based routing protocol for underwater sensor networks. Sensors. 2012;**12**(4):4559-4577. DOI: 10.3390/s120404559

[30] Bilbao J, Crespo PM, Armendariz I, Médard M. Network coding in the link layer for reliable narrowband powerline communications. IEEE Journal on Selected Areas in Communications. 2016;**34**(7):1965-1977. DOI: 10.1109/JSAC.2016.2566058

[31] Douik A, Sorour S, Al-Naffouri TY, Yang HC, Alouini MS. Delay reduction in multi-hop device-to-device communication using network coding. In: International Symposium on Network Coding (NetCod); 22-24 June 2015. IEEE; 2015. pp. 6-10

[32] Zhan C, Xu Y. Broadcast scheduling based on network coding in time critical wireless networks. In: IEEE International Symposium on Network Coding (NetCod); 9-11 June 2010. IEEE; 2010. pp. 1-6

[33] Lun DS, Médard M, Koetter R, Effros M. On coding for reliable communication over packet networks. Physical Communication. 2008;**1**(1):3-20. DOI: 10.1016/j.phycom.2008.01.006

[34] Li S, Maddah-Ali MA, Avestimehr AS. Coding for distributed fog computing. IEEE Communications Magazine. 2017;**55**(4):34-40. DOI: 10.1109/MCOM.2017.1600894

[35] Marques B, Machado I, Sena A, Castro MC. A communication protocol for fog computing based on network coding applied to wireless sensors. In: International Symposium on Computer Architecture and High Performance Computing Workshops (SBAC-PADW). IEEE; 2017. pp. 109-114

[36] Sipos M, Fitzek FHP, Lucani D, Pedersen M. Dynamic allocation and efficient distribution of data among multiple clouds using network coding. In: 3rd International Conference on Cloud Networking (CloudNet); 8-10 October 2014. IEEE; 2014. pp. 90-95

[37] Cabrera Guerrero J, Lucani D, Fitzek F. On network coded distributed storage: How to repair in a fog of unreliable peers. In: International Symposium on Wireless Communication Systems (ISWCS); 20-23 September 2016. IEEE; 2016. pp. 188-193

[38] Dimakis AG et al. Network coding for distributed storage systems. Transactions on Information Theory. IEEE. 2010;**56**(9):4539-4551. DOI: 10.1109/CloudNet.2014.6968974

[39] Sipos M, Heide J, Lucani D, Pedersen M, Fitzek F, Charaf H. Adaptive network coded clouds: High speed downloads and cost-effective version control. IEEE Transactions on Cloud Computing. 2015;**99**:1-1. DOI: 10.1109/TCC.2015.2481433

[40] Oliveira PF, Lima L, Vinhoza TTV, Barros J, Medard M. Coding for trusted storage in untrusted networks. IEEE Transactions on Information Forensics and Security. 2012;7(6):1890-1899. DOI: 10.1109/TIFS.2012.2217331

[41] Ostovari P, Wu J. Fault-tolerant and secure distributed data storage using random linear network coding. In: 14th International Symposium on Modeling and Optimization in Mobile, Ad Hoc, and Wireless Networks (WiOpt); 9-13 May 2016. IEEE; 2016. pp. 1-8

Efficient Frontier and Benchmarking Models for Energy Multicast in Wireless Network Coding

Adeyemi Abel Ajibesin

Additional information is available at the end of the chapter

http://dx.doi.org/10.5772/intechopen.79377

Abstract

This chapter introduces efficiency frontier and benchmarking concepts to evaluate the efficiency performance of wireless networks for multicast energy. These concepts are efficiency models based on the data envelopment analysis (DEA) technique. The DEA framework allows network administrators to evaluate the technical efficiency and determine how the inefficient wireless networks will attain a targeted efficiency frontier. In order to achieve efficiency frontier and benchmark by a wireless network, this chapter presents several models including the envelopment and the slack. The envelopment model evaluates the technical efficiency scores of each wireless network, while the slack model shows how the inefficient wireless network achieves efficiency frontier. The benchmark model evaluates the efficiency reference set and the lambda values of each network. The efficiency frontier algorithm has shown that many of the wireless networks sampled are inefficient. However, the algorithm has capability to help the inefficient wireless networks to achieve efficiency frontier and benchmark with their peers that are fully efficient.

Keywords: efficiency frontier, network coding, modeling, wireless networks, multicast energy

1. Introduction

Technical efficiency evaluation and expectation are new kinds of thinking for many evaluators especially in the field of network coding [1, 2]. The current approach to coded packet evaluations is largely dependent on average measurement [3]. This type of approach is only good to demonstrate the impact of a program but inadequate to evaluate the technical efficiency and benchmark [4]. One of the major factors in the evaluation of efficiency is the limited resources

IntechOpen

and decisions on how to allocate such resources. This requires a special consideration in evaluation processes [5, 6].

In literature, the essence of minimum energy multicast is to optimize high-energy transmission over the network. This was achieved using the minimum energy multicast algorithm. However, the minimum energy multicast problem is NP-hard [7]. The alternative solutions using polynomial time-based heuristics approach were considered [8–10]. One of these solutions is the multicast incremental power algorithm. As an improvement to this technique, the minimum energy multicast problem in ad hoc wireless networks is solvable as a linear program, assuming network coding technique [11]. Compared with conventional routing solutions, network coding technique does not only promise a potentially lower multicast energy but also enables finding the optimal solution in polynomial time. Other energy efficiency algorithms presented in the literature for energy efficiency were all designed to achieve similar goals using the effective performance evaluation approach [12–14].

In this chapter, a network coding algorithm is studied and its performance is investigated for the data evaluation analysis (DEA) technique. The DEA methodology is necessary because the coded packet is not a fully efficient technique for energy efficiency [15]. The DEA, which was used to study the relative efficiency and productivity of systems in economic and operational research (OR) disciplines, is a nonparametric method that relies on linear programming techniques for optimizing discrete units of observation called the decision-making units (DMUs) [16]. The DEA method is different from other because it adopts the frontier analysis approach to evaluate efficiency rather than averages and standard deviation [16]. Therefore, our system model is based on frontier analysis that consists of several models including envelopment and benchmarking. These models are considered for evaluating the technical efficiency of multicast energy and performing the benchmark in wireless network nodes without affecting the overall network performance.

The remainder of this chapter is presented in sections: Section 2 provides necessary background information on the minimum energy multicast and Section 3 presents the network coding performance that is based on average multicar energy. Section 4 and 5 discuss the efficiency frontier method and benchmarking model, respectively. In Section 6, efficiency frontier implementation and results analyzed are discussed while Section 7 concludes the chapter.

2. Background

This section begins with the discussion of energy-efficient multicast following the various multicasting techniques used to minimize the wireless multicast energy.

2.1. Energy-efficient multicast

Researchers have worked on energy-efficient networking for several years especially with the growth of the wireless networks such as wireless sensor networks, mesh networks, and ad hoc wireless networks. Many studies have explored the topic of energy efficiency of these networks [17–19]. Some of the studies that were investigated in literature include routing, coding, cross-layer designs, MAC protocols, spectrum allocation, resource allocation, and scheduling. The scope of this chapter is to present the actual efficiency of multicast energy in wireless networks. So it discusses energy efficiency in the routing. An approach to energy efficiency is the exploration of the broadcast nature of the wireless links. Wireless links are either omnidirectional or directed over a large area to ensure that transmissions are received by more than one node. This feature has effects on multicast networks, and it is known as wireless multicast advantage (WMA) [20, 21]. In routing, the problem of performing energy-efficient multicast considering WMA is NP-complete [22]. Thus the problem of minimum energy broadcast and multicast is solved in wire-line cases by various minimum weight-spanning tree algorithms but the solutions are generally suboptimal [23]. However, alternative approach using the network coding method was employed [24–27].

2.2. Minimum-energy multicast

The main optimization problem for energy efficiency broadcast and multicast routing in ad hoc wireless networks is to minimize the total transmission power assigned to all nodes [15]. This is widely recognized as one of the performance challenges in wireless networking. The minimum energy multicast problem in ad hoc wireless networks is solvable using several approaches. A popular approach is the minimum shortest path tree (MSPT) algorithm that has been applied to solve minimum energy network problems [22]. This algorithm builds minimum energy networks and measures the cost (energy) of an edge based on certain levels [23]. However, this problem is known to be NP-hard [7]. An alternative approach such as minimum spanning tree that is based on the greedy heuristic algorithm was proposed [9]. The method used can compute minimum energy in polynomial time, thereby reducing the cost (energy) on multicast tree twice than that of SMPT. However, the solutions provided by this approach are suboptimal. In order to improve the solutions, a large number of approximation algorithms were proposed for energy-efficient multicast in wireless networks including a unique method to improve the energy efficiency of multicast trees using pruned or greedy heuristics [21]. In the literature, the performances of three greedy heuristics algorithms, multicast incremental power (MIP) algorithm, multicast least-unicast-cost (MLU) algorithm, and multicast link-based MST (MLiMST) algorithm, were analyzed [22]. It has been shown that MIP algorithm has best performance for all network nodes that are considered. However, the MIP approach is also suboptimal. Thus, the network coding technique has been considered for improved performance [25].

3. Network coding performance

In this section, the performance of a network coded algorithm is investigated and the results serve as imputes for the frontier analysis. We consider a flow-based approach that addresses networks with costs such as energy using a linear programming technique [26]. The cost is a function of coding subgraph z. We represent the cost function with ξ. This approach assumes that all nodes in the network are capable of coding with a focus on the problem of minimizing network resource such as multicast energy. We represent this function with z which is the coding subgraph. We then consider a formulated multicast problem connection, which is a triplet $(S, T, \{R_t\})_{t \in T}$, where S is the source of the connection, T is the set of network receivers, and R_t is the set of rates at which the flow is being injected to the sinks. Furthermore, the multicast connections using the random linear network coding (RLNC) algorithm that has been proved in the literature to address such problems are considered. The optimization formulation for this problem is given as:

$$\min \xi(z)$$

subject to

$$z \in Z$$

$$\sum_{j \in K} x^t_{(iJj)} \leq z_{iJK} b_{iJK}, \quad \forall (i, J) \in H, \quad K \subset J, t \in T, x^t \in F^t$$

$$\min \xi(z)$$

subject to

$$z \in Z$$

$$\sum_{j \in J} x^t_{(iJj)} \leq z_{iJ}, \quad \forall (i, J) \in H, \quad t \in T, x^t \in F^t$$

where $x^t_{(iJj)}$ represents the average rate of the packets that are injected on the hyper arc link and received by exactly the set of nodes J, which occurs with the average rate z_{iJ} and that allocated to a particular connection. F^t is the bounded polyhedron of points

x^t satisfying the conservation of flow constraints. We consider a lossless network with multicast applications and made some assumptions [27]. For example, it is assumed that when nodes transmit, they reach all other nodes in certain regions, with cost increasing as the region expands. These assumptions have helped the problem to reduce in the case of linear separable cost and separable constraints. Therefore, a fixed cost such as energy can easily be evaluated while the constraints set for Z are dropped. Readers are referred to [28] for more details about this formation. A well-known RLNC algorithm, which is appropriate to deploy network coding in a real multicast network, is considered for the simulation of this optimization problem. The details and the pseudocode for the RLNC algorithm are presented in [29].

The authors have considered various network parameters which include the network sizes, the radius of connectivity, the dimension for the nodes, the source nodes, and the receiving nodes. Randomly generated nodes were simulated and the average energy of the multicast networks was evaluated using the RLNC algorithm. The effective performance of the network coding algorithm presented has shown the limitation of the algorithm and the evaluation approach in terms of efficiency performance [30]. It is important to understand that "effectiveness" is mainly concerned with achieving a set goal. For instance, network effectiveness is the ability of such a network to attain its predetermined goals. For instance, one of the goals of the RLNC algorithm is to minimize energy such that the results or outcome is better than the previous algorithms [31, 32]. This evaluation approach is concerned on the right way of minimizing multicast energy rather than how well the multicast energy is being minimized. Thus, efficiency is concerned with how well the multicast energy is minimized. This is achieved by quantitatively evaluating the ratio of output to input. With efficiency evaluation, the performance is based on the combination of both inputs and outputs rather than focusing on the outcome results (outputs) only. For example, in [33], the results presented based on average performance show that the performance of the RLNC algorithm in minimizing energy was degraded as the number of sinks increased but improved as the network size increased. This result has been shown to perform better than the existing algorithm when compared. However, it is an effective performance and cannot determine the efficiency of the algorithms.

4. Efficient frontier method

Data envelopment analysis (DEA) is a nonparametric method that relies on the linear programming technique for optimization using frontier analysis. It is used to measure the relative efficiency of peer decision-making units (DMUs) that have multiple inputs and outputs [34]. Unlike network coding evaluation method that is based on average performance evaluation, the frontier method is used to evaluate the technical efficiency of DMUs. Besides, the efficiency frontier technique is capable of improving the input resources as well augment the output results while the performance remained the same. In case of input resources, the multicast energy of a wireless network is considered to be minimized, while the number of sinks remained the same. Also, in the case of output augmentation, the number of sinks can be increased, while the multicast energy is kept constant. Furthermore, the efficiency frontier method evaluates the performance of a wireless network by comparing its efficiency with the best observed performance in the data set. Thus, efficiency frontier represents the best observed performance among the networks [35].

4.1. Illustrating efficient frontier

This concept of efficiency frontier is best explained with a simple case of one input and one output. Let us consider the data in **Table 1** where the technical efficiency of each set of eight wireless networks (DMUs) is evaluated. A data value for each DMU is provided. We plot the data in **Table 1** with input on the x axis (the horizontal axis) and the output on the y axis (the vertical axis) to obtain **Figure 1**. This figure shows the technical efficiency of each DMU. The figure also shows the picture of efficient frontier. The wireless network with efficiency frontier is the one that floats on top of data observations.

Figure 1 shows the wireless network E on the efficiency frontier with an efficiency of 1. The line that spans from the origin through the wireless network (DMU E) is known as efficiency

DMU	A	B	C	D	E	F	G	H
Output	4	6	6	8	10	10	12	16
Input	2	3	4	6	10	4	6	10

Table 1. Data of simple efficiency ratio to evaluate efficient frontier.

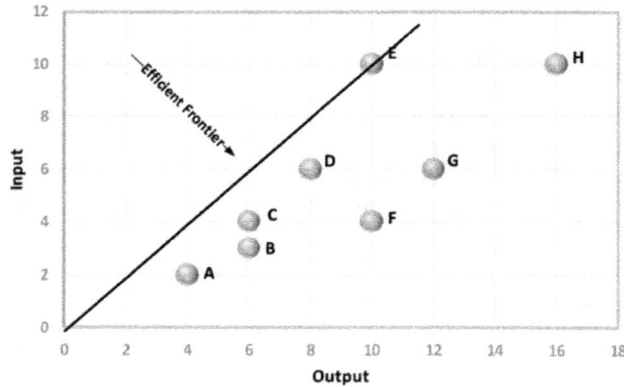

Figure 1. DMU on efficient frontier versus inefficient DMUs.

frontier [36]. The inefficient wireless networks are located beneath the efficient frontier. These inefficient wireless networks can be moved unto the efficient frontier using an orientation approach [36]. There are two fundamental directions to achieve this move: The input-oriented and the output-oriented approaches. The input-oriented approach will be applied to reduce the multicast energy while the number of receives is fixed at their current levels. The output-oriented approach is outside the scope of this chapter. Using input orientation and considering the wireless network F, which is an example of inefficient network, a projection to point E can be performed. This is the targeted position for wireless network F to become efficient. In the real world, most problems are multidimensional in nature with many input and output variables. As a result, the efficiency frontier using DEA solver that is based on the linear programming technique is considered for the evaluation of efficiency frontier in this chapter.

4.2. Efficient frontier system model and procedure

A wireless network (DMU) that lies on the efficiency frontier is said to attain its targeted energy level. The main problem that this chapter addresses is that many networks multicast their messages using average energy rather than targeted energy. A wireless network administrator, especially at this stage of technological development, cannot base network evaluation on average performance. Therefore, one of the problems is that given the different set of wireless networks with a node (source) multicast to some selected group of nodes (receivers) using average energy, how can we qualitatively evaluate performance so that they attain targeted energy? This problem is impossible to answer without the efficiency frontier method.

The existing approach was to calculate the average energy multicast and then rank them according to the lowest. The lowest average energy multicast is considered the most effective network. However, the lowest average energy multicast does not mean it is the most efficient [37]. We state that any wireless network that multicasts messages to a selected group of nodes using targeted or projected energy is said to attain efficiency frontier. Performance according to the efficiency frontier is possible if a network makes use of the combination of its multiple inputs and multiple output resources correctly.

Figure 2 presents the flowchart that is used to solve this problem. The flowchart consists of different steps. The first step, which is envelopment model, evaluates the technical efficiency scores of a wireless network. Subsequently, the second step, which is the slack model calculates and classifies the efficient wireless into full or weak networks. The last step is the projection model that determines how the weakly efficient wireless network will be fully efficient so that they also attain efficiency frontier. These procedures are computed using the DEA solver and the efficiency frontier results are compared with the average energy computed using the network coding algorithm. The differences in multicast energy are recorded. If there is no difference, it means that the average energy used by RLNC is fully efficient. Otherwise, it is

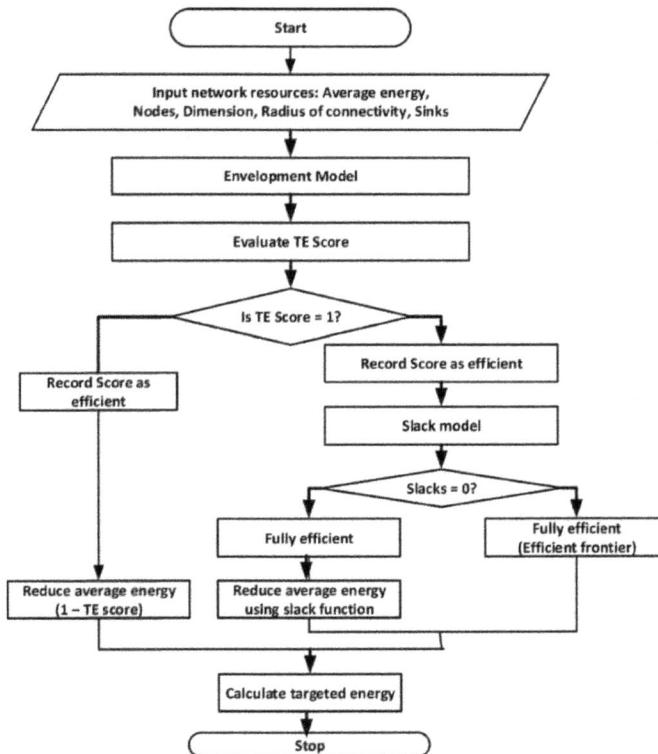

Figure 2. Algorithm of the targeted multicast energy based on efficiency frontier approach.

inefficient or weakly efficient. However, as discussed, the efficiency frontier method provides mechanisms for making the inefficient wireless network to achieve efficiency. The models considered are based on Charnes, Cooper, and Rhodes (CCR) with assumption of constant returns to scale (CRS) [38].

4.3. The envelopment model

This chapter considers the minimization of multicast energy using efficiency frontier method that relies on the linear programing (LP) technique of the DEA. The LP is an approach to evaluate a set of weights that yields the maximum efficiency. An appropriate envelopment DEA model that evaluates energy efficiency was presented in [39] and is given below:

$$\theta^* = min \ \theta$$

subject to

$$\sum_{j=1}^{n} \lambda_j x_{ij} \leq \theta x_{i0}, \qquad i = 1, 2, ..., m;$$

$$\sum_{j=1}^{n} \lambda_j y_{rj} \geq y_{r0}, \qquad r = 1, 2, ..., s; \tag{1}$$

$$\lambda_j \geq 0, \qquad j = 1, 2, ..., n,$$

where λ_j are unknown weights with $j = 1, 2, ..., n$ and they correspond to the *DMU* numbers. DMU$_0$ is one of the n DMUs under evaluation, and θx_{i0} and y_{r0} are the ith input and rth output for DMU$_{0,}$ respectively.

The following conditions are required for the calculation of efficiency scores: If $\theta^* = 1$, then the DMU under evaluation is a frontier point (fully or weakly efficient). Otherwise if $\theta^* < 1$, then the DMU under evaluation is inefficient. To address inefficiency, the DMU can either increase its output levels or decrease its input levels to achieve efficiency [40]. The θ^* represents the efficiency score of *DMU$_o$* based on input-orientation. This means that the model is able to minimize energy while maintaining the current output levels.

4.4. The slack model

The slack model is needed to push the weak efficient or inefficient wireless networks to their real efficiency frontier so that targeted energy is achieved. The linear programming formulated for slack model is given as [40, 41]:

$$max \ \sum_{i=1}^{m} s_i^- + \sum_{r=1}^{s} s_r^+$$

subject to

$$\sum_{j=1}^{n} \lambda_j x_{ij} + s_j^- = \theta^* x_{io}, \qquad i = 1, 2, \ldots, m;$$

$$\sum_{j=1}^{n} \lambda_j y_{rj} - s_r^+ = y_{r0}, \qquad r = 1, 2, \ldots, s; \tag{2}$$

$$j \geq 0 \qquad\qquad j = 1, 2, \ldots, n$$

where s_j^- and s_r^+ represent input and output slacks, respectively. The superscripts $(-)$ and $(+)$ represent input reduction and output augmentation, respectively. The condition for fully (100%) efficient is if and only if both (a) $\theta^* = 1$ and (b) all slacks $s_i^{(-*)} = s_r^{(+*)} = 0$. The targeted multicast energy can be calculated using the following expressions:

$$\begin{cases} X_{i0}^* = \theta^* x_{io} - S_i^{-*}, & i = 1, 2, \ldots, m \\ Y_{r0}^* = y_{ro} + S_r^{+*}, & r = 1, 2, \ldots, s \end{cases} \tag{3}$$

This is calculated by multiplying the average multicast energy with an optimal efficiency score (θ^*), and slack amounts are subtracted.

5. Benchmarking model

In this section, a variable-benchmark model is considered for minimum energy multicast. The variable benchmark allows a new wireless network to be evaluated against a set of given benchmarks or standards. Also, it is formulated upon input-oriented CCR/CRS model. The model extends the envelopment and slack models discussed in the previous section. The benchmark model determined the efficiency reference set (ERS) and the amount required by each wireless network to catch up with their peers. The remainder of this section presents the mathematical function and the requirements for benchmark evaluation.

In the process of developing a benchmark, once the efficiency frontier is established, we can compare a set of new wireless networks with the reference efficiency frontier. The idea is that whenever a new wireless network outperforms the identified efficiency frontier, a new efficiency frontier is generated by the DEA solver. This means that the benchmark for a wireless network is different from other new wireless networks depending on network condition and variables used. The benchmark model contributes to how a wireless network learns the best way to utilize the available resources [42]. The benchmark model first evaluates the efficiency reference set (ERS) and the amount required by each wireless network to catch up with their peers. This magnitude is called the lambdas.

In order to formulate variable benchmark, the envelopment model is modified for the benchmark optimization problem as follows:

$$\textit{Minimise } \alpha^{CCR/CRS}$$

subject to

$$\sum_{j\in E^*} \lambda_j x_{ij} \leq \alpha^{CCR/CRS} x_i^{new}$$

$$\sum_{j\in E^*} \lambda_j y_{rj} \geq y_r^{new} \qquad\qquad (4)$$

$$\lambda_j \geq 0, j \in E^*,$$

where $\alpha^{CCR/CRS}$ represents the optimal value to model, and E^* represents the set of benchmarks identified by the DEA. The new observation is represented by DMU^{new} with inputs $x_i^{new}(i = 1, 2, ..., m)$ and outputs $y_r^{new}(r = 1, 2, ..., s)$. The superscript of CCR/CRS indicates that the benchmark composed by benchmark DMUs in set E^* is based on CCR/CRS model. Model represents the performance of DMU^{new} with respect to benchmark DMUs in set E^*, when outputs are fixed at their current levels. Furthermore, model is capable of yielding a benchmark for DMU^{new}. Thus the ith input and the rth output for the benchmark can be expressed as:

$$\begin{cases} \sum_{j\in E^*} \lambda_j^* x_{ij} & (ith\ input) \\ \\ \sum_{j\in E^*} \lambda_j^* y_{rj} & (rth\ ouput) \end{cases} \qquad (5)$$

The expression (5) indicates that although the DMUs associated with set E^* are given, the resulting benchmark may be different for each new DMU under evaluation. Thus, there is a variable-benchmark scenario.

6. Implementation, results, and discussions

This section begins with brief overview of the software used for the implementation of the algorithm presented in Section 4 and 5. It then discusses and analyses the results obtained from the models.

6.1. DEA solver for efficient frontier analysis

The frontier analysis is evaluated using the DEA software, which is the tool that was specially packaged to solve the envelopment model and other types of DEA models. The efficiency frontier analysis relies on the DEA library, which includes the Solver and LPsolver (linear programming solver) program to perform optimizations. This work makes use of DEAOS for the implementation of the efficiency frontier models. The DEAOS is a web-based software. The readers are referred to [43] for details about the DEAOS package and user's documentation. The DEA implementation procedures were discussed in [40].

6.2. Technical efficiency performance

The DEA solver compares each DMU with all other DMUs and identifies those DMUs that are operating inefficiently. It also evaluates the magnitude of inefficiency of the DMUs that are suboptimal. The efficient DMUs are those that attain efficient frontier and are identified by a DEA efficiency rating of $\theta = 1$. The inefficient DMUs are identified by the efficiency score of less than 1 ($\theta < 1$). Column 1 of **Table 2** is the results of the average multicast energy computed by the RLNC reports. This result was presented in [43]. Column 2 of **Tables 2** and **3** report the results of DEA technical efficiency and inefficiency scores of 54 wireless networks, respectively. From **Table 2**, only DMU_9, DMU_{18}, DMU_{27}, and DMU_{45} have the efficiency score of $\theta = 1$ (i.e., 100%) and thus they are identified as efficient. Other DMUs have efficiency scores of less than 1 ($\theta < 1$) but greater than 0 and are identified as inefficient. It is possible for inefficient DMUs to improve their technical efficiency scores by reducing certain inputs using input orientation. For example, DMU_1 can improve its technical efficiency score by reducing certain inputs up to 73.4% (100−26.6). Similarly, DMU_2 can do so with approximately 63.1% of input reduction. However, DMU_{36} is closer to an efficiency frontier and needs only a 2.4% reduction of its input resources. This is achieved using the slack model.

6.3. Evaluation of slacks and targeted multicast energy

Column 4 of **Table 2** presents the targeted results using slack and projection. In the slack model, none of the efficient DMUs have a slack, meaning that slacks exist only for those DMUs identified as inefficient. The slacks are obtained after proportional reductions in inputs. The slack is essential whenever a wireless network cannot reach the targeted multicast energy. Then, slacks are required to project such wireless networks to the targeted multicast energy which is their efficient frontier. The general rule is that a DMU with at least a slack input value is needed to be projected into the frontier, but a DMU that has zero slack for all the inputs does not need any projection because it already reached targeted efficient frontier. The targeted multicast energy is calculated by multiplying the average multicast energy with the technical efficiency score, and the slack values are subtracted. This calculation is used to achieve the target set for multicast energy.

6.4. Benchmarking for ERS and lambdas evaluation

The benchmark model addresses the benchmark problem. It is a model for establishing the standard of excellence. The model is able to determine the efficiency reference set (ERS) and lambdas of the inefficient wireless networks. Lambdas define the amount of inputs to be reduced for an inefficient wireless network to catch up with their peers that are already operating efficiently. We consider the same data set used for envelopment and slack model. The implementation procedures for benchmarking are also similar. The same DEA solver is

DMU	Ave. energy (RLNC)	Efficiency score (%)	Inefficiency score (%)	Targeted energy
DMU_1	4.5	30	70	1.3
DMU_2	5.5	40.1	59.9	2.2
DMU_3	6.2	49.2	50.8	3.1
DMU_4	6.8	58	42	4
DMU_5	7.3	66.3	33.7	4.9
DMU_6	7.2	78	22	5.6
DMU_7	8.1	82.4	17.6	6.7
DMU_8	8.8	90	10	7.6
DMU_9	8.5	100	0	8.5
DMU_{10}	5.2	27.5	72.5	1.4
DMU_{11}	5.6	39.5	60.5	2.2
DMU_{12}	6.3	48.7	51.3	3.1
DMU_{13}	6.9	57.3	42.7	3.9
DMU_{14}	7.1	67.1	32.9	4.8
DMU_{15}	7.2	77.8	22.2	5.6
DMU_{16}	7.7	84.4	15.6	6.5
DMU_{17}	8.6	90	10	7.5
DMU_{18}	8.3	100	0	8.3
DMU_{19}	4.2	30.3	69.7	1.3
DMU_{20}	5.3	36.5	63.5	1.9
DMU_{21}	5.4	48.3	51.7	2.6
DMU_{22}	6.1	54.5	45.5	3.3
DMU_{23}	6.2	64.5	35.5	4
DMU_{24}	6.4	73.4	26.6	4.7
DMU_{25}	6.6	81.8	18.2	5.4
DMU_{26}	7.3	90	10	6.1
DMU_{27}	6.7	100	0	6.7
DMU_{28}	3.6	34.9	65.1	1.3
DMU_{29}	5.1	37.7	62.3	1.9
DMU_{30}	5.6	46.8	53.2	2.6
DMU_{31}	5.9	56.1	43.9	3.3
DMU_{32}	6.1	65.3	34.7	4
DMU_{33}	6.8	69.9	30.1	4.7
DMU_{34}	6.6	81.2	18.8	5.4
DMU_{35}	7.1	87	13	6.2
DMU_{36}	7.1	96.8	3.2	6.9

DMU	Ave. energy (RLNC)	Efficiency score (%)	Inefficiency score (%)	Targeted energy
DMU$_{37}$	3.1	40.1	59.9	1.3
DMU$_{38}$	4.6	41	59	1.9
DMU$_{39}$	4.8	53	47	2.5
DMU$_{40}$	4.8	66.2	33.8	3.2
DMU$_{41}$	5.6	68	32	3.8
DMU$_{42}$	5.6	79	21	4.4
DMU$_{43}$	6.3	80.6	19.4	5
DMU$_{44}$	6.3	90.1	9.9	5.7
DMU$_{45}$	6.3	100	0	6.3
DMU$_{46}$	3.6	34.8	65.2	1.3
DMU$_{47}$	4.3	43.8	56.2	1.9
DMU$_{48}$	5.1	49.7	50.3	2.5
DMU$_{49}$	5.1	61.5	38.5	3.2
DMU$_{50}$	5.5	69.3	30.7	3.8
DMU$_{51}$	5.7	76.8	23.2	4.4
DMU$_{52}$	6.4	78.7	21.3	5.1
DMU$_{53}$	6.4	88.6	11.4	5.7
DMU$_{54}$	6.5	97.6	2.4	6.3

Table 2. Results of the average multicast energy computed by network coding (RLNC) algorithm, the envelopment model, (efficiency and inefficiency), and the projected multicast energy computed by DEA Solver.

used for the benchmark model. The benchmark model is able to identify the ERS and calculate the lambda values.

Table 3 is extracted from the DEA simulation output sheet. The network administrators whose network is inefficient can observe the benchmark networks that they need to catch up with. From **Table 3**, the full efficient network may consider itself to be its own "benchmarks." This is because efficient network has already achieved 100% efficiency. So, benchmark for DMU9 is DMU9 and for DMU18 is DMU18. The same applies to DMU27 and DMU45. However, for inefficient ad hoc networks, their benchmarks are one or many of the efficient ad hoc networks. For example, a benchmark for DMU2 and DMU3 are DMU9, DMU18 and DMU27. This means, DMU2 and DMU3 must use a combination from DMU9, DMU18 and DMU27 to become efficient.

Another benchmark analysis is the lambda value. This benchmark analysis calculates the amounts of benchmark needed from a DMU to achieve efficiency. These values are reported as magnitude (lambda) next to each benchmark DMU on **Table 3**. For instance, as seen from **Table 3** and as shown in **Figure 3**, DMU16 will attempt to become like DMU18 (blue bar) more than DMU27 (red bar) as observed from their respective lambda weights of DMU18 and DMU27 ($\lambda 18 = 71.3$ and $\lambda 27 = 8.7$).

DMUs	Efficiency reference set (ERS)	Lambdas values (%)		
DMU_1	DMU18, DMU27	0.020	19.98	
DMU_2	DMU9, DMU18, DMU27	7.310	2.520	20.17
DMU_3	DMU9, DMU18, DMU27	19.28	2.300	18.41
DMU_4	DMU9, DMU18, DMU27	32.03	2.000	15.98
DMU_5	DMU9, DMU18, DMU27	45.89	1.570	12.54
DMU_6	DMU9, DMU18, DMU27	51.99	2.000	16.00
DMU_7	DMU9, DMU18, DMU27	74.54	0.610	4.850
DMU_8	DMU9	90.00		
DMU_9	DMU9	100.0		
DMU_{10}	DMU18, DMU27	5.000	15.00	
DMU_{11}	DMU18, DMU27	10.91	19.09	
DMU_{12}	DMU18, DMU27	22.69	17.31	
DMU_{13}	DMU18, DMU27	35.30	14.70	
DMU_{14}	DMU18, DMU27	45.86	14.14	
DMU_{15}	DMU18, DMU27	54.47	15.53	
DMU_{16}	DMU18, DMU27	71.28	8.720	
DMU_{17}	DMU9	90.00		
DMU_{18}	DMU18	100.0		
DMU_{19}	DMU45	20.00		
DMU_{20}	DMU27, DMU45	10.44	19.56	
DMU_{21}	DMU27, DMU45	15.14	24.86	
DMU_{22}	DMU27, DMU45	36.41	13.59	
DMU_{23}	DMU27, DMU45	46.63	13.37	
DMU_{24}	DMU27, DMU45	59.82	10.18	
DMU_{25}	DMU27, DMU45	74.70	5.300	
DMU_{26}	DMU27, DMU45	31.69	58.31	
DMU_{27}	DMU27	100.0		
DMU_{28}	DMU45	20.00		
DMU_{29}	DMU27, DMU45	6.970	23.03	
DMU_{30}	DMU27, DMU45	19.50	20.50	
DMU_{31}	DMU27, DMU45	31.77	18.23	
DMU_{32}	DMU27, DMU45	44.20	15.80	
DMU_{33}	DMU18, DMU27	0.360	69.64	
DMU_{34}	DMU27, DMU45	76.40	3.600	
DMU_{35}	DMU18, DMU27	8.970	81.03	
DMU_{36}	DMU18, DMU27	9.610	90.39	
DMU_{37}	DMU45	20.00		

DMUs	Efficiency reference set (ERS)	Lambdas values (%)	
DMU_{38}	DMU45	30.00	
DMU_{39}	DMU45	40.00	
DMU_{40}	DMU45	50.00	
DMU_{41}	DMU45	60.00	
DMU_{42}	DMU45	70.00	
DMU_{43}	DMU45	80.00	
DMU_{44}	DMU45	90.00	
DMU_{45}	DMU45	100.00	
DMU_{46}	DMU45	20.00	
DMU_{47}	DMU45	30.00	
DMU_{48}	DMU45	40.00	
DMU_{49}	DMU45	50.00	
DMU_{50}	DMU45	60.00	
DMU_{51}	DMU45	70.00	
DMU_{52}	DMU27, DMU45	5.290	74.71
DMU_{53}	DMU27, DMU45	5.630	84.37
DMU_{54}	DMU27, DMU45	9.660	90.34

Table 3. ERS and lambdas of input-oriented variable benchmark.

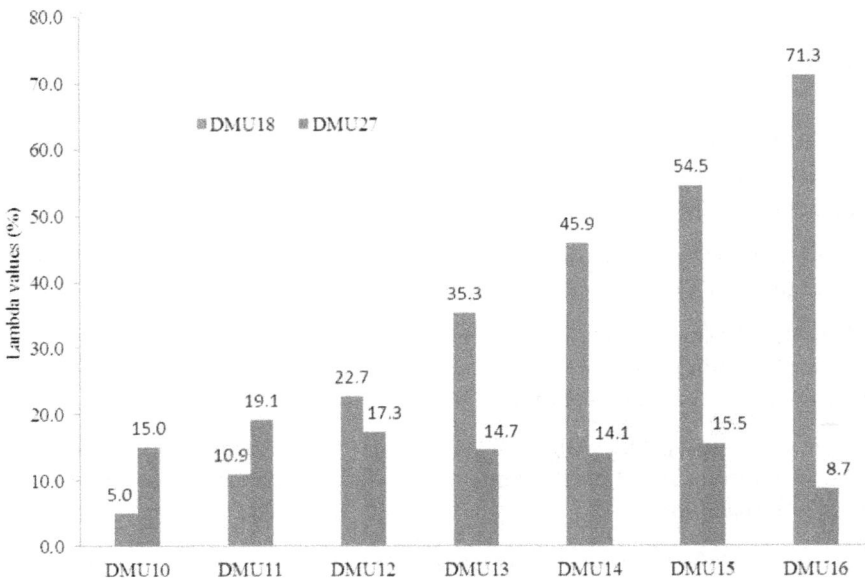

Figure 3. Benchmarks and lambdas of the input-oriented variable benchmark.

7. Conclusion

This chapter studied the existing network coding algorithm and investigated the efficiency performance of the multicast energy in wireless networks. The previous reports have shown that network coding based on effective evaluation is sub-optimal because they were largely calculated using central tendency performance such as average and standard deviation. While effective performance is a good evaluation tool, it is not enough to measure the actual efficiency of networks. In order to appropriately evaluate the network efficiency, a new algorithm based on efficiency frontier was considered for the evaluation. With this approach, the targeted multicast energy for wireless networks is achieved using envelopment, slack, and benchmarking models. These models were formulated upon input-oriented CCR/CRS assumptions. The aim of this chapter was to achieve economic efficiency by ensuring that wireless networks are multicast at the targeted energy rather than average energy. Furthermore, this was achieved without sacrificing the network performance.

Author details

Adeyemi Abel Ajibesin

Address all correspondence to: abel.ajibesin@aun.edu.ng

School of Information Technology and Computing, American University of Nigeria, Yola, Nigeria

References

[1] Hassan Mohammed A, Dai B, Huang B, Azhar M, Xu G, Qin P, Yu S. A survey and tutorial of wireless relay network protocols based on network coding. Journal of Network and Computer Applications. Mar. 2013;36(2):593-610

[2] Minn J, Zeng H, Vijay K. Green cellular networks: A survey, some research issues and challenges. IEEE Communication Surveys and Tutorials. 2012;13(4):524-540

[3] Lilien LT, Othmane LB, Pelin A, DeCarlo A, Salih RM, Bhargava B. A simulation study of ad hoc networking of UAVs with opportunistic resource utilization networks. Journal of Network and Computer Applications. Feb. 2014;38:3-15

[4] Begum IA, Buysse J, Alam MJ, Van Huylenbroeck G. Technical, allocative and economic efficiency of commercial poultry farms in Bangladesh. World's Poultry Science Journal. Aug. 2010;66(03):465-476

[5] Raayatpanah MA, Salehi Fathabadi H, Khalaj BH, Khodayifar S, Pardalos PM. Bounds on end-to-end statistical delay and jitter in multiple multicast coded packet networks. Journal of Network and Computer Applications. May 2014;41:217-227

[6] Katrina P. Understanding Cost-Effectiveness of Energy Efficiency Programs: Best Practices, Technical Methods, and Emerging Issues for Policy-Makers. A Resource of the National Action Plan for Energy Efficiency; 2008

[7] Cagalj M, Hubaux JP, Enz C. Minimum-energy broadcast in all-wireless networks: NP-completeness and distribution issues. In: ACM MobiCom; 2002. pp. 172-182

[8] Guo S, Yang O. Minimum-energy multicast in wireless ad hoc networks with adaptive antennas: MILP formulations and heuristic algorithms. IEEE Transactions on Mobile Computing. 2006;**5**(5):333-346

[9] Yuan D, Bauer J, Haugland D. Minimum-energy broadcast and multicast in wireless networks: An integer programming approach and improved heuristic algorithms. Ad Hoc Networks. Jul. 2008;**6**(5):696-717

[10] Wieselthier EJ, Nguyen GD, Anthony E. Algorithms for energy-efficient multicasting in static ad hoc wireless networks. Mobile Networks and Applications. 2001;**6**(3):251-263

[11] Ho T, Lun DS. Network Coding: An Introduction. Cambridge, UK: Cambridge University Press; 2008

[12] Ajibesin AA, Ventura N, Murgu A, Chan HA. Energy minimization in WSNs: Empirical study of multicast incremental power algorithm. In: South Africa Telecommunication Networks and Applications Conference (SATNAC); Port Elizabeth, South Africa; 31 August–3 September, 2014

[13] Mueen U, Azizah AR. Virtualization implementation model for cost effective and efficient data centers. International Journal of Advanced Computer Science and Applications. 2011;**2**(1):59-68

[14] Wen Y-F, Liao W. Minimum power multicast algorithms for wireless networks with a Lagrangian relaxation approach. Wireless Networks. Jun. 2011;**17**(6):1401-1421

[15] Ajibesin AA, Ventura N, Murgu A, Chan H. Data envelopment analysis: Efficient technique for measuring performance of wireless network coding protocols. In: 15th International Conference on Advanced Communication Technology; 2013. pp. 1122-1127

[16] Cooper WW, Lawrence MS, Kaoru T. Data Envelopment Analysis: A Comprehensive Text with Models, Applications, References and DEA-Solver Software. New York, USA: Springer Science + Business Media; 2007

[17] Blume O, Zeller D, Barth U. Approaches to energy efficient wireless access networks. In: 4th International Symposium on Communications, Control, and Signal Processing; 2010

[18] Lin R, Wang Z, Sun Y. Energy efficient medium access control protocols for wireless sensor networks and its state-of-art. In: 2004 IEEE International Symposium on Industrial Electronics. Vol. 1; 2004

[19] Louhi J. Energy efficiency of modern cellular base stations. In: 29th International Telecommunications Energy Conference (INTELEC); 2007. pp. 475-476

[20] Etoh M, Ohya T, Nakayama Y. Energy consumption issues on mobile network systems. In: International Symposium on Applications and the Internet (SAINT 2008); 2008

[21] Zhang H, Gladisch A, Pickavet M, Tao Z, Mohr W. Energy efficiency in communications. IEEE Communications Magazine. 2010;**48**(11):48-49

[22] Akbari Torkestani J, Meybodi MR. A link stability-based multicast routing protocol for wireless mobile ad hoc networks. Journal of Network and Computer Applications. Jul. 2011;**34**(4):1429-1440

[23] Lun DS, Ratnakar N, Medard M, Koetter R, Karger DR, Ho T, Ahmed E. Minimum-cost multicast over coded packet networks. IEEE Transactions on Information Theory. Jun. 2006;**52**(6):2608-2623

[24] Wieselthier JE, Nguyen GD, Ephremides A. Energy-efficient broadcast and multicast trees in wireless networks. Mobile Networks and Applications. 2002;**7**(6):481-492

[25] Biradar R, Manvi C, Sunilkumar S. Review of multicast routing mechanisms in mobile ad hoc networks. Journal of Network and Computer Applications. Jan. 2012;**35**(1):221-239

[26] Wu M, Kim C. A cost matrix agent for shortest path routing in ad hoc networks. Journal of Network and Computer Applications. Nov. 2010;**33**(6):646-652

[27] Liang W. Approximate minimum-energy multicasting in wireless ad hoc networks. IEEE Transactions on Mobile Computing. 2006;**5**(4):377-387

[28] Eslami A, Khalaj BH. Capacity of network coding for wireless multicasting. In: IEEE Annual Wireless and Microwave Technology Conference; 2006. pp. 1-5

[29] Ho T, Medard M, Koetter R, Karger DR, Effros M, Shi J, Leong B. A random linear network coding approach to multicast. IEEE Transactions on Information Theory. Oct. 2006;**52**(10): 4413-4430

[30] Ajibesin AA, Wajiga GM, Odekunle MR, Egunsola OK. Energy-efficient for multicast networks: A new approach to efficiency measure. In: 8th IEEE EUROSIM Congress on Modelling and Simulation (EUROSIM2013); Cardiff, Wales, United Kingdom; September 9–12, 2013. pp. 616-621

[31] Fragouli C, Soljanin E. Network coding applications. Foundation and Trends in Networking. 2007;**2**(2):135-269

[32] Li Z, Li B. Network coding in undirected networks. Computer (Long Beach, California). 2004:1-6

[33] Ajibesin AA, Ventura N, Chan HA, Murgu A, Egunsola OK. Performance of multicast algorithms over coded packet wireless networks. In: 2012 UKSim–14th International Conference on Modelling and Simulation; Cambridge: Cambridgeshire United Kingdom; March 28–30, 2012

[34] Cooper WW, Lawrence MS, Joe Z. Handbook on Data Envelopment Analysis. 2nd ed. New York, USA: Springer Science + Business Media; 2011

[35] Kastaniotis G, Maragos E, Douligeris C, Despotis DK. Using data envelopment analysis to evaluate the efficiency of web caching object replacement strategies. Journal of Network and Computer Applications. Mar. 2012;**35**(2):803-817

[36] Coelli TJ, Rao DS, O'Donnell CJ, Battese GE. An Introduction to Efficiency and Productivity Analysis. 2nd ed. New York, USA: Springer Science + Business Media; 2005

[37] Katrina P. The Measurement of Productive Efficiency and Productivity Growth (Google eBook). Oxford University Press; 2008

[38] Banker WW, Charnes RD, Cooper A, et al. Management Science. 1984;**30**(9):1078-1092

[39] Cullinane K, Song D-W, Wang T. The application of mathematical programming approaches to estimating container port production efficiency. Journal of Productivity Analysis. Sept. 2005;**24**(1):73-92

[40] Ajibesin AA, Ventura N, Chan A, Murgu A. DEA envelopment with slacks model for energy efficient multicast over coded packet wireless networks. IET Journal of Science, Measurement and Technology. 2014;**8**(6):408-419

[41] Tone K. A slacks-based measure of efficiency in data envelopment analysis. European Journal of Operational Research. 2001;**130**(3):498-509

[42] Joe Z. Qualitative Models for Performance Evaluation and Benchmarking: Data Envelopment Analysis with Spread Sheets. New York, USA: Springer Science + Business Media LLC; 2009

[43] Data Envelopment Analysis. 2015. [Online]. Available from: http://www.deaos.com/ [Accessed: May 30, 2018]

www.ingramcontent.com/pod-product-compliance
Lightning Source LLC
Chambersburg PA
CBHW081242190326
41458CB00016B/5890